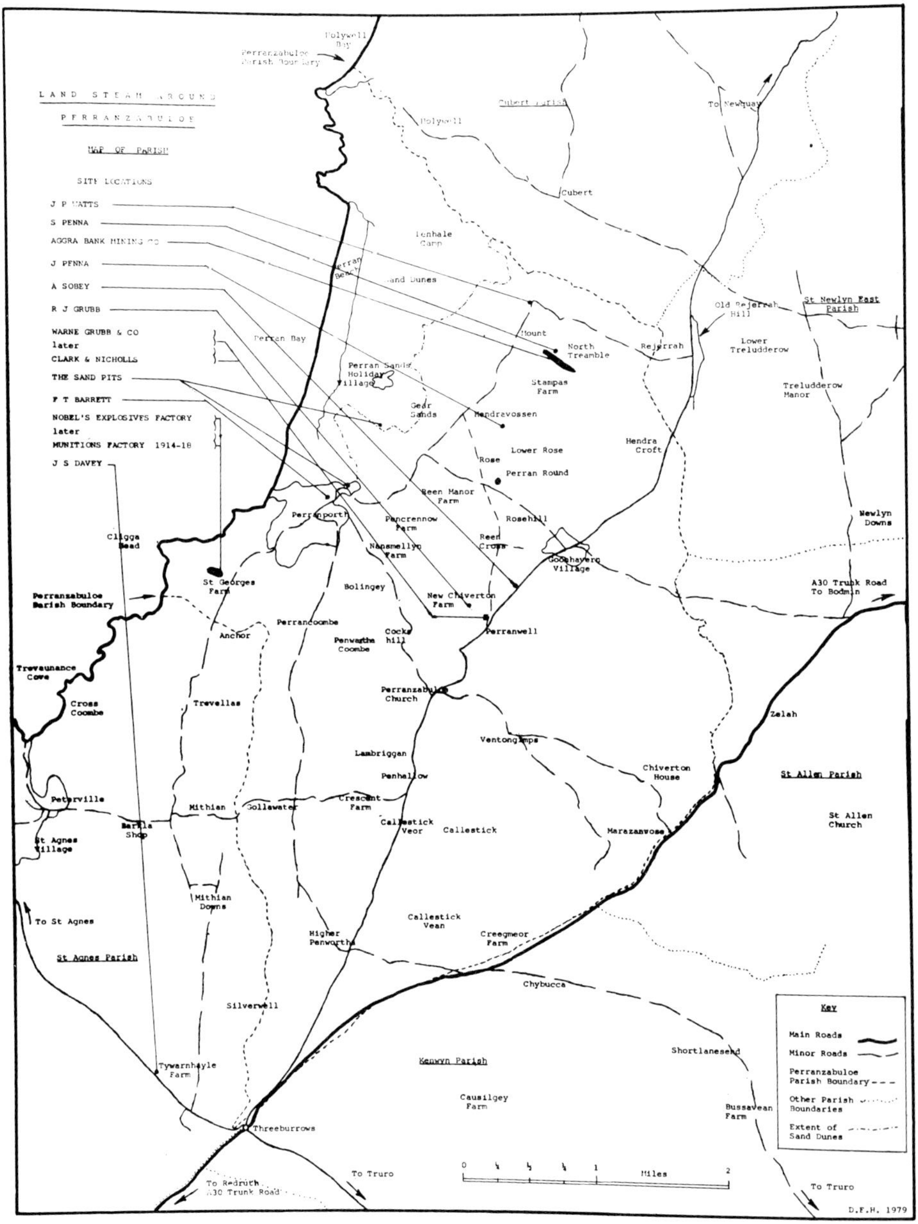
LAND STEAM AROUND
PERRANZABULOE
MAP OF PARISH
SITE LOCATIONS
J P WATTS
S PENNA
AGGRA BANK MINING CO
J PENNA
A SOBEY
R J GRUBB
WARNE GRUBB & CO
later
CLARK & NICHOLLS
THE SAND PITS
F T BARRETT
NOBEL'S EXPLOSIVES FACTORY
later
MUNITIONS FACTORY 1914-18
J S DAVEY
Holywell Bay
Perranzabuloe Parish Boundary
Cubert Parish
Holywell
To Newquay
Cubert
Penhale Camp
Perran Beach
Sand Dunes
Old Rejerrah Hill
St Newlyn East Parish
Mount
North Treamble
Rejerrah
Lower Treludderow
Perran Bay
Perran Sands Holiday Village
Stampas Farm
Treludderow Manor
Gear Sands
Hendravossen
Hendra Croft
Lower Rose
Rose
Perran Round
Been Manor Farm
Cligga Head
Perranporth
Pencrennow Farm
Rosehill
Newlyn Downs
Reen Cross
Nansmellyn Farm
Goonhavern Village
St Georges Farm
Bolingey
A30 Trunk Road To Bodmin
Perranzabuloe Parish Boundary
New Chiverton Farm
Perrancoombe
Perranwell
Anchor
Cocks hill
Penwartha Coombe
Trevaunance Cove
Perranzabuloe Church
Cross Coombe
Trevellas
Zelah
Ventongimps
Lambriggan
Chiverton House
St Allen Parish
Penhallow
Peterville
Mithian
Gollawater
Crescent Farm
St Allen Church
Barkla Shop
Callestick Veor
Callestick
Marazanvose
St Agnes Village
Mithian Downs
To St Agnes
Callestick Vean
Higher Penwartha
Creegmeor Farm
St Agnes Parish
Chybucca
Silverwell
Key
Main Roads
Minor Roads
Perranzabuloe Parish Boundary
Other Parish Boundaries
Extent of Sand Dunes
Tywarnhayle Farm
Kenwyn Parish
Shortlanesend
Causilgey Farm
Bussavean Farm
Threeburrows
0 ¼ ½ ¾ 1 Miles 2
To Redruth A30 Trunk Road
To Truro
To Truro
D.F.H. 1979

CORNISH LAND STEAM
in and around Perran Parish

D.E. Hattam.

First published in 1983
by Dyllansow Truran
Trewolsta, Trewirgie, Redruth, Kernow
Typeset in Great Britain by Delta Graphics, Falmouth, Cornwall
Designed and Printed by Penryn Litho, Mabe, Penryn, Cornwall

ISBN 0 907566 63 4

Contents

It is said, Cornwall was the cradle of the steam engine ....

At a time when steam was being abandoned in favour of alternatives, a few devotees, ignoring the financial aspect, kept steam alive in Cornwall by preserving and working their traction engines. In so doing, they maintained an unbroken record of well over 250 years usage of steam. Few, if any, places can boast of a longer record. It is to these men and women that this book is dedicated.

Introduction

This work covers a period of over eighty years, tracing the earliest conventional traction engines in Cornwall up to the first engine in preservation. It is centred on the parish of Perranzabuloe and, it is hoped, will be of interest to the steam enthusiast, local historian and holiday maker alike.

The parish of Perranzabuloe is one of the bigger parishes of Cornwall and is situated on the north coast. In the 1870's it was comprised of some 10,955 acres, with a scattered population of around 3,600 but steadily falling. At that time the industries were mining and agriculture with a little fishing. Today, agriculture is still dominant, mining and fishing are extinct and tourism is now a major factor, especially on the coastal side.

The highways and roads of those days were a far cry from the relatively wide, straight, rebuilt smooth surfaced roads we have today. They were narrow, steep and twisting, rarely wide enough for two vehicles to pass each other and often had ruts, cut by regular traffic, over a foot deep. Constructed of macadam, in summer many were dust tracks and in winter, pools of mud and slime. It was not until the 1920's that they were reconstructed.

Over these roads in all winds and weathers, men plied their engines and loads, sometimes for mining and others for agricultural purposes. On the land, from late summer to early spring, they worked with their machines, threshing from farm to farm, often walking or cycling miles before and after work. These men accepted their lot and got the job done — it was their living.

In a little work such as this it is not possible to list all those people who have provided information. Over fifty people have received me courteously and answered my queries to the best of their ability. Some information has been provided by searches in the contemporary local press and other journals. Without both sources this work would not have been possible. To all concerned I say thank you. The staff of Royal Institute of Cornwall, Redruth Reference Library and County Record Office must also be thanked as also, Shirley Force who read the manuscript.

My notes are available for inspection by anyone sufficiently interested.

Derek E. Hattam St. Agnes December, 1979

J. Sydney Davey, Tywarnhayle Farm, St. Agnes

Joshua Sydney Davey was the second son of Stephen Davey of Bochym, Cury, near Helston. The Davey family were also vested with the Manors of Mithian and Tywarnhayle, both in St. Agnes parish. In 1869, Sydney Davey, an energetic man who liked things done properly, was twenty-seven years old and lived at Town Farm, Redruth. He was already an established speculator in Cornish mines, manager of Redruth Brewery and also ran a model farm at Tywarnhayle. Much of this farm had been broken in by his father and uncle and now comprised some two hundred acres of arable land and over a hundred acres of grass pasture. He had won several cups for the fine management and breeding on the farm. (The farm has now been broken up. It was situated adjacent to the road running from the A30 trunk road to St. Agnes, now known as Teagle straight). Most of the fields ran into several acres, the average being about 25; large by general Cornish standards.

On 13th September 1869, the Steam Plough Works of John Fowler & Co. Leeds, completed building and despatched by rail, two ploughing engines, together with various implements, to Mr. J.S. Davey, Redruth. These engines were works numbers 1187 and 1188.

On Monday 20th September they arrived at Truro Station and created a considerable sensation. During the evening steam was raised, they were unloaded with their implements and then driven by road to Tywarnhayle Farm. The engines were described as splendid specimens of workmanship and they created an immense commotion as they proceeded down Richmond Hill and up Kenwyn Street, the noise attracting thousands of people to witness the novel sight! They left Truro by way of the steep Chapel Hill, which was easily ascended at 3 m.p.h., onto the Redruth road, "going on their journey without difficulty". The engines were of 12 nominal horse power and described, rather innacurately, as weighing 10 tons. A complete set of implements had been purchased and included a five furrow antibalance plough, an eleven tine cultivator, a harrow, a threshing machine and a winnowing machine. The complete tackle had cost the owner more than two thousand pounds.

Next day they were set to work ploughing and again on

Wednesday, at Mr. Davey's farm "doing the work exceedingly well". These engines were the first tractions to be introduced into Cornwall for agricultural purposes and, therefore, attracted a great number of spectators, many of them farmers. The engines generally caused great excitement in the neighbourhood and daily "large numbers came to witness the operations, when not less than ten acres were ploughed". Apparently, it was Mr. Davey's policy to accommodate the needs of neighbouring farmers who chose to use the tackle. There was considerable praise from the press for his enterprising spirit and aspersions cast on the rich land owners for not having exhibited similar enterprise.

In the next week, during three days, forty acres of stiff land were cultivated to an average depth of nine inches, at Tywarnhayle Farm, and by the end of the week, the tackle had been engaged by several farmers, some of whose land was situated on moderate slopes. (The operation known as cultivating is called chiselling in Cornwall). It was considered that twenty horses could not have worked the same land in the same time and also thought that on average land, fifteen acres a day could be ploughed five to six inches deep, twenty acres a day could be cultivated, or, forty to fifty acres harrowed. Fields down to eight acres could be readily worked but smaller fields would be charged at the eight acre rate. The charge for ploughing was set at ten shillings (50p) an acre and for cultivating eight shillings (40p) an acre. The cost was considered to be about equal to horse labour but advantages were to be gained by way of getting more ploughing done in good weather and, also, increased productivity of the land. One man was employed on each engine with another on the plough or cultivator and the farmer provided "boys or girls to attend to the pulleys". It was said that although the tackle would dispense with horses it did not dispense with manual labour and throw workmen out of work.

As might be expected, the use of children around such machinery eventually led to disaster. Before the end of October, while the tackle was threshing at a Mr. Tiddy's farm, near Goonvrea, a boy named John Williams, eight years of age, was standing at the end of the machine, when a large barrel used for carrying water accidentally fell over and caught the boy's head between the barrel and machine causing instant death.

On Tuesday 2nd November, the engines with all the tackle proceeded to Treswithian Farm, Camborne, where they were to be employed by Mr. Alfred Edmunds. On their way they passed through Redruth town, approaching it from the north and "the bottom of Fore Street was all astir about two o'clock in the afternoon". At the bottom of the town they would be turning west, heading for Camborne. In those days, Chapel Street was much narrower where it joins Fore Street, a substantial shop having been demolished to make way for modern traffic. "The engines came round a sharp corner from the north road, in splendid style, attracting the attention of some hundreds of persons". At Camborne, they "attracted a large concourse of people to review the machine in transit". The crowd were not to be disappointed; "Two powerful traction engines — their tall chimneys puffing out volumes of smoke and steam — were seen coming down the street, having attached to them all the appurtenances necessary for ploughing and cultivation of the land by steam".

The tackle was at work next morning. At an early hour, each engine "steamed into its respective position and cultivated in a most superior style two statute acres per hour, to a depth of from eight to nine inches". After a field of eight acres had been cultivated, about thirty gentlemen, guests of Mr. Edmunds, were entertained at a lavish lunch, in a tent erected in the field. For some reason the party was presided over by the rector of Camborne, supported by the rector of Redruth and of course, Sydney Davey. Mr. Davey's health was drunk and praises made concerning his public spirit and scientific farming, the style of which "was never before practiced in the West of England". In response, he stated that the tackle had cost more than two thousand pounds and he was determined, if possible, to make it pay and also be very useful. In the evening the party had tea at the farm house when several of those present expressed their approval of the steam plough, believing it to be a step in the right direction.

On the Saturday evening of that week, Sydney Davey read a short paper at Camborne to members of the East Penwith Agricultural Exchange, the subject being steam cultivation and entitled "Steam versus Horse Power in Agriculture". The paper was described by the press as "able" and it is included here, ad verbatum, being as interesting in a different way to the reader with hindsight, as it must have been to his audience with foresight.

'Mr. President and Gentlemen — you have asked me to bring before you a few remarks respecting "Steam Cultivation". I am sorry that I have been called upon thus early to give an opinion, because my practical experience is but short, and must necessarily be imperfect. However, I beg to draw your attention to a short paper, viz. "Steam versus Horse Power".

'As steam-power for agricultural purposes is not generally developed in this county (Cornwall), although we have it to perfection, or very near it, for mining purposes, my object at present is to show you the advantages of steam power connected with agriculture. We all know, or ought to know, the cost of agricultural horses. Many of them, probably, are not properly kept, but that is on the "penny wise pound foolish principle". There is not the least doubt that two well fed horses are equal in strength to three neglected horses. At the same time, as we keep horses for the sake of "power", we must feed them with corn, hay, chaff, & co., to get the power which is produced by the muscle and flesh, and after that power is obtained in an ox or a horse, how is it to be employed or made available but by yoking it to an agricultural implement? Therefore, in taking a field which is rather foul (I mean a crop of stroil or couch grass) we put our quadrupeds into a cultivator or plough, or any other implement which, in the first place, have got up their power by hay or corn, which food ought to be made into beef and mutton instead of horse flesh, which is not saleable in this country for food, except for kennels and that at a very low rate. Again, after the animals are yoked, they must tread on the ground very firm to move the implement, and thereby so poach the land that we are giving ourselves double labour (especially if the soil is in any way inclined to a clayey nature) either in ploughing or cultivation; the furrow is raised in a firm lump, and there it remains until the spring, when it requires no end of harrowing and rolling at the time we ought to have everything ready for our seed sowing without delay, especially in our uncertain Cornish climate. We must also take into consideration our casualties of horses, the depreciation of value, the wear and tear, and the interest of money on our general agricultural implements, also a certain system that our ploughmen and boys have in "Touching a pipe", and when you come up they say, "Oh! If you please sir, we're obliged to put on a new share or a point"; and then the horses are shivering by the hedgeside, when they often catch a chill if sweating after ploughing hard and deep land, thence we are sometimes minus a horse that has died of inflammation or something else.

'Now in regard to steam cultivation, we have the advantage of working our land whenever the weather is most suitable, which is a great benefit, as one dry and suitable day and a large quantity of work done is very valuable, and gives satisfaction to the farmer, as it saves a deal of labour in the spring or any other time, for this reason:- the engines being locomotive (Messrs. Fowler's), are fixed on each headland and propel themselves by the hedgerow, the plough of five farrows or the cultivator of eleven tines works backwards or forwards between each engine, drawn by a steel rope; therefore, when the steam is up each person has his own proper employment and works regularly, and no time is lost; the soil is also left in a very light and aerated state, and not trod down as when done with horses, which advantage is that in the spring we just put the harrows on and then there is a perfect seed bed.

'Probably there may not be any remarkable difference in the expense of steam power or horse power in Cornwall for ploughing or cultivating, as ours is naturally light land, and seldom ploughed more than four or five inches deep, but it tells a tale in favour of steam in the spring, when we have the largest portion of our crops to sow, and in a moist climate such as we have here, with little time to do it in, especially if the land has been deeply cultivated in the autumn.

'There can be no doubt of the advantages of two engines instead of one engine and an anchor as formerly used. More work can be done in a day, and, at the same time, more effectual; they take half the quantity of rope, which is a very expensive item, and are more convenient for transit. From reports, the single engine and anchor improved the crops from ten to twenty per cent, and I am confident that by the double engine set, where the work is done more effectually, we may expect a greater increase in our crops.

'Steam is now universally used for thrashing and other farm work, and the advantage of a traction engine is shown by the machine being taken by itself instead of sending horses for it. I am satisfied that ultimately they will come more into general use. The expense is heavy, but I have every hope that the speculation I have embarked in will prove satisfactory, and that after my friends have seen the results of the first years working they will come to me and give me full employment.

'I thank you gentlemen for the kind way in which you have listened to me, and trust that I have brought before you a few remarks for discussion.'

The usual discussion took place afterwards and several members, who had seen the quality of the work at Treswithian, were well pleased with it. The engines were still at work in the Camborne area at the end of the month, being requested by many farmers who were "delighted with its performances".

A letter at this time, in the correspondence columns of the newspapers, reveals that all was not quite as well as we may expect. "Hence, if you discuss in many of our rural districts the pros and cons of the newly introduced steam plough, it stands immediately condemned". The reason given stated that deep ploughing was condusive to the growth of charlock (a weed), which was, in fact, pure fallacy. An invitation by the anonymous correspondent to discuss the matter, failed to attract interest, although an editorial article the following week favoured the tackle and quoted a letter from an experienced agriculturalist, farming a large estate in west Cornwall. In it he stated that, having seen the tackle at work, it was his opinion that the sooner steam cultivation came into the County the better, and on a large farm in the spring, he would prefer it to thirty horses, feeling sure that it would do more work and be more efficient. He wished there were ten such enterprising men as Sidney Davey in the county of Cornwall. Another article, strangely, thought deep ploughing would improve fox hunting!

Discussions continued during the winter months, the topic being frequently brought up at the traditional dinners following local ploughing matches. Mr Davey of Crafthole, an implement maker, speaking to Roseland District Agricultural Improvement Society, held the belief that steam ploughing would never be of great value to Cornwall, due to the smallness of the fields and hilly nature of the land, making it difficult to operate. Nevertheless, speakers in other parts of the county accepted that people in the Western District appeared to speak very highly of the steam plough.

During January 1870, a large stretch of unbroken, open downs, belonging to the Tehidy Estate, skirting the North Cliffs between Spratin Cove and Hells Mouth, was burnt in preparation for cultivation. The fire lasted two days. On the sixteenth of March the plough was set to work on this common land. It proved difficult, being full of furze (gorse) and other deep rooted bracken, but some two hundred acres were taken in by mid May. The tackle was then moved, by road, to Launceston for the Royal Cornwall Show. By then, in nine months, some seven hundred acres of land had been ploughed, two hundred of which had been waste land. In practice it had been found that on average, eight acres a day were ploughed, fifteen acres cultivated and thirty acres could be harrowed.

At Launceston, it was the most interesting exhibit in the implement section, and, in spite of the fact that it was working about a mile from the main show ground, it brought large numbers of visitors to see it at work. Most of them were practical men, who were pleased with its successful working. The field in which the tackle was working was on the top of a hill and the engines could not be seen from each other. The plough was set at seven inches and the cultivator twelve; the cultivator "came up the hill at more than five miles an hour and only two minutes were occupied turning it at the headland". It was stated that the tackle was the only one in the whole of the four Western Counties and, that Mr. Davey had not only recouped his capital, but also made a handsome profit. Whilst this may have been so, it is difficult to see how.

Working ten hour days, six days a week, in nine months there would be two hundred and thirty four days available. Seven hundred acres, ploughed at an average of eight acres a day and priced at ten shillings an acre would gross £440. Assuming that the rest of the days were spent cultivating or harrowing another gross £876 would be brought in. From the total of £1,316 expenses for wages, coal, oils, etc., have to be deducted. The calculation does not allow for any idle days during the working week. The rate for breaking in the two hundred acres at Tehidy estate was no doubt higher than normal but probably would not have amounted to over £1,400 to bring the accounts anywhere near balance point.

After the show, the activities of the tackle seem to have been taken more for granted, the novelty, no doubt, having worn off. In January 1871 Mr. Davey was advocating the cultivation of wastelands by steam power and believed it would pay. Later that year, he stated that he did not have enough time to supervise the steam tackle and suggested that a company be formed to run it.

An examination of account books show that during the year some £2,083. 5. 2d. was credited to the "Steam Tackle" account. The accounts are difficult to follow, as virtually no credits are shown for other years before or after this date. On the debit side, the engines appear to have required very little maintenance. The accounts do not show entries for things like coal, plough shares etc., so cannot be wholly relied on. They continue until October 1875 but details are vague. Nevertheless, they indicate that the tackle was still in his possession at that time. About then, Mr. Davey moved to Bochym, near Helston and an account of that place, written some years later, does not make any mention of the engines so we must assume that they did not move with him.

What then were these engines like? By researching various records, drawings, writings and photographs it is possible to build up an accurate description of the engines.

Photographs of these particular engines (numbers 1187 &1188), rebuilt on identical boilers by the Oxford Steam Plough Co., are still in existence. From an inspection of these photographs it is obvious from the original parts that the engines were of the vertical drive (shaft to drum) type and single cylindered. The photographs show Allen type safety valves, long smoke box and no governor on the rebuilds.

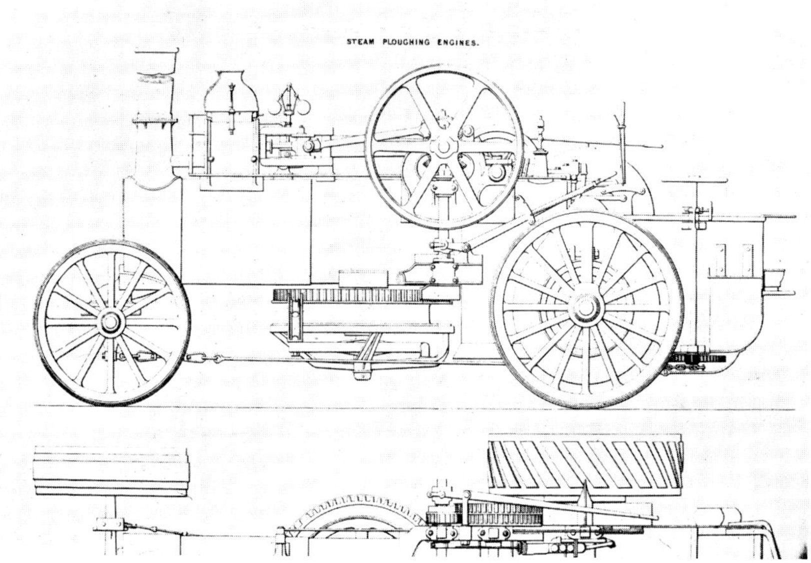

Line drawing of Fowler ploughing engine as supplied to J.S. Davey in 1869. (The plan view at the bottom shows the clutch and gearing). From Minutes of proceedings of the Institute of Civil Engineers, Vol XXXVI, (1872—3, Part II), page 69 et seq.

During 1869, Fowlers started introducing this type of engine, eventually to supersede their already developed horizontal shaft type. These engines, therefore, must have been amongst the very earliest of what was to become the conventional design for ploughing engines. Later types only varied in detail and size.

Originally, the engines were built on a long loco-type boiler, 2′ 9″ diameter, without hornplates. It was fitted with 29 tubes, 2¾″ diameter, and had a grate area of 7.6 square feet and total heating surface of 201 square feet, pressurised for working up to 150 p.s.i. The ash pan damper was on the off side. The crankshaft, dished fly wheel and bevel gear was carried on a pair of massive, heavy castings, bolted to the boiler. This shaft crossed the boiler just in front of the firebox to allow the bevelled gear of the vertical shaft, mounted on the near side, to drive the winding drum, held by wrought iron stud fixed to the bottom of the boiler. The shaft pinion meshed with an externally toothed gear ring on the drum, which was fitted with a brake and carried 400 yards of rope and self-acting coiling gear. It was engaged by a clutch above the lower bearing of the vertical shaft, (in much the same manner as on the later BB1's). The crank was kept close to the bevel bearing which determined that the steam chest be on the off side.

The steam jacketted cylinder diameter was 10½″, with a 12″ stroke and produced up to 103 I.H.P.. Stephensons link reversing gear was fitted, operated from the offside of the footplate, with the link down in forward gear (i.e. rods crossed — not ideal for economy of steam). The cross head ran in angle iron guides. A Watt-type governor was mounted on the weigh shaft bracket and a single Salter safety valve, set to lift at 150 p.s.i., was anchored to the top of the cylinder casting, the valve itself being contained within a dome. The regulator was operated by a nearly horizontal unequal lever, mounted across the boiler and carried forward along the nearside to an unequal rocking lever connecting it with the valve rod which entered the back end of the block. Cylinder drains were now operated from the footplate. A whistle was mounted over the firebox end of the boiler as was a boiler pump on the offside. This was driven by an eccentric on the crank shaft and the pump was inclined upward to meet it. This was the only feedwater supply.

Set close to the bearing on the gear side of the crankshaft, the low gear pinion was fitted to run loosely on the shaft outside the bearing and in mesh at all times. A dog clutch, driven on the squared end of the crankshaft and operated by a long lever to the footplate, engaged with the pinion for low gear work (1½ m.p.h.). For faster work, at 3 m.p.h., a detachable pinion took the place of the clutch which could also be removed. This pinion engaged with a second spur wheel bolted to the larger spur on the second shaft. When not in use, the detachable pinion and clutch was carried in a box on the front axle. The bearings for these shafts were carried in the heavy casting already mentioned, as was the top bearing for the vertical shaft. All the gearing was of cast steel.

Another pair of large castings carried both the third shaft and hind axle. The third shaft transmitted the drive to the other side of the engine and down to the axle onto which friction drums were keyed, one on each side, inside the wheels. The drive was then transmitted to the wheels with the use of friction bands, which tightened on the drums. The band was anchored to a very wide spoke deeply embedded in the cast boss of the wheels which were not fitted with hubcaps. The wrought iron wheels were 5′ 6″ diameter, 20″ wide and fitted with strakes which were inclined upwards toward the outer edge of the wheel, when viewed from the rear. The strakes also appear to have been fitted more acutely to the edge of the wheel than we are used to. The engines measured 8′ 2½″ across the wheels. It will be realised that the friction bands were used instead of drive pins. Unfortunately, they had a habit of occasionally slipping on steep hills!

The tender was fairly low and straight-backed. It carried 180 gallons of water and 8 cwts of coal. The water pocket was on the back and two separate wrought iron bars, one above the other, provided the central drawing hitch. Steps to the doorless footplate were on the offside. All the steering gear was mounted on the tender. The steering wheel was situated behind the nearside hind wheel and operated from a step on top of a toolbox at the corner of the tender. To turn the engine right, the wheel had to be turned left! At the bottom of a vertical shaft from this wheel, a pinion was keyed and meshed with a spur gear under the tender. Bolted to this gear and concentric with it, was a sprocket which meshed with a chain passing around it and carried forward to link with long rods. These in turn were connected, by adjustable links, to brackets extended down from the front axle, which swivelled on a flat turntable. The steersman was provided with a friction brake under the steering wheel, tightened by a small wheel, to hold the steerage on rough roads . An obstruction in front of one of the wheels could spin the wheel out of his hands or throw him off his step, because the steering was so directly geared.

A wooden toolbox was mounted in front of and carried on the front axle, which had 4′ 6″ diameter wheels. This axle was set well forward of a very short smokebox, with flat door, by an extended perch bracket. Rubbing plates were fitted to protect the smokebox from the front wheels which had hub caps. A tall stack base, carried a very tall, slightly tapered stack with plain top. The engines did not have foot boards along the boiler. The oval maker's plate and number was placed halfway along the boiler (nearside), which had four belly bands. A large boiler door was cut in the top of the barrel.

The engines weighed 13½ tons and were 21′ 6″ long. To the eye, the combination of extended perch bracket, low single cylinder, low tender and tall slim stack, gave an overall impression of a long low engine, compared to the later BB1.

Exactly what happened to the tackle after 1875 is not known but the engines eventually found their way to Kent, where they had several different owners. They were reboilered by John Allen & Sons in the early part of the century and they later carried the registration numbers KE 4285 and KE 4284. They were still in existence circa 1950, since when they have been cut up for scrap.

Note: Examples of early 12 H.P. single cylinder Fowler ploughing engines have survived for preservation. They have, however, all been reboilered and all have had many detail modifications made at various times. The nearest engine to those described is number 1190, reboilered by Wilder's and fitted with new cylinder, modified perch bracket, smoke box and clutch gear. The other, somewhat similar engines, are numbers 1368, 2013, 2479 and 2693. The latter are considerably modified.

Although Davey is here credited with introducing steam ploughing to the West Country, this is not strictly true. Trevithick had tried a steam plough of sorts in Cornwall in 1804 and a Mr. H. Parsons of Crewkerne, Somerset, used a steam plough in 1862 on the roundabout system. It is, however, quite true that Davey introduced steam ploughing, on the double engine system, to the West Country (i.e. Cornwall, Devon, Somerset and Dorset).

For some reason, steam ploughing did not catch on in Cornwall. No doubt major factors against it were the generally small fields, by "up country" standards, which would have been wasteful in leaving headlands unploughed or to be ploughed by horses. The undulating countryside and narrow gateways to the fields must have been other factors, quite apart from the narrow twisting roads and lanes bordered by thick, sometimes tall, hedges. The gateways could have been modified but not the roads and lanes.

At this time also, great strides were being made in the design of horse ploughs and the reputation of the leading implement maker in Cornwall, J. Davey of Crafthole, was second to none. By coincidence or intention, he introduced his balance and turnwrest horse ploughs in quick succession following the arrival of the steam plough.

Finally, the stubborn and thrifty nature of the Cornish farmers, resisting change and capital outlay, must have played, perhaps, the major part in the supremity of the horse plough in Cornwall for well over another half century; by which time steam was being overtaken by oil engines.

Only one other ploughing engine came new to Cornwall.

Agra Bank Mining Co.

Running inland from the northern end of Perran Bay, across the parish of Perranzabuloe, is a large iron ore deposit, known as the Great Perran Iron Lode. In the early seventies of the last century there was a shortage of iron ore and deposits all over the country were being tried. The presence of this ore body had been known for a long time but due to its proximity to established ports, a long haul across country was necessary before its shipment to Wales for smelting. One of the schemes considered long ago was the building of a harbour and jetties at Holywell Bay. Quite rightly this idea was abandoned in favour of using more natural ports. The railway system at this time had not spread around the County and the sudden need for the now valuable iron ore prompted several Bills to Parliament for an Act to allow certain railway lines to be laid, connecting the mines to the ports. In the meantime, however, work started on the mines and the ore had to be transported to Truro for shipment. The roads of that time were not much more than cart tracks and the motive energy came from horses and mules. It is not surprising, therefore, that the coming of a traction engine drew the attention of the contemporary mining and local press. Indeed, this was one of the earliest Road engines in Cornwall and the first in Perranzabuloe.

In February 1872, reports appeared in the local papers and Mining Periodicals of a mammoth mining operation to win iron ore from the Great Perran Iron Lode. The works were to be of a character "to astonish the natives", and a capital outlay of a million pounds — a considerable sum in those days — was envisaged. The work was scheduled to commence in March, carried out by the contractor, Sir Morton Peto. At the same time application had been made for the laying of a mineral railway line. Pending the completion of this line efforts were subsequently made to convey the ore to its place of shipment, Truro, by a traction engine. A contemporary report in the Mining Journal observed that "there are few Counties in England in which the difficulties of working these engines are greater than in Cornwall and therefore the attempt must be looked upon as experimental only". The route was in fact difficult, traversing among others the steep valleys of Rejerrah and Perranwell. At that time the only experience of traction haulage on roads in Cornwall was with the ploughing tackle of S. Davey which "proved that his engines will master most of the hills with their ploughing gear", but another observer pointed out "it would require a heavier load of ore to make their employment pay".

With the purchase of an engine already completed, by the Agra Bank Mining Co., Messrs Farley & Co., of Falmouth, were instructed to manufacture ten trucks, each to carry five tons. In addition to transporting iron ore, they were also to be used for carrying coal to the various mines in Perranzabuloe on the return trips. The shipment of ore was planned to take place from Garras Wharf, Truro, or if impractical there, from Malpas on the Truro river.

On Monday 11th March 1872, the traction engine arrived by rail at Truro station and on Tuesday "a sensation was created in the town by its being driven through the streets". Proceeding from the top of Richmond Hill, "on its way down it went gently along and the novelty of the sight attracted a large number of persons out of doors to witness it, and many who had not the opportunity of seeing it were alarmed by a rumour that an engine had broken loose from the railway station and was running about the town, an absurdity which many appeared to believe".

The driver kept his pressure low and its progress through the town was successful until it reached the neighbourhood of the Cornish Bank (near the War Memorial) where "the incompetency of the persons managing it first began to appear, and the behaviour of the machine afterwards was such as to give many persons the impression that it was an utter failure". Apparently the driver had let his water get low and left the fire untended so that the plan to take the engine down Malpas road had to be abandoned. "A slight accident to the fore wheel marred the success of the experiment". According to another report of a more optimistic author, "there is little doubt it will do its work well". The engine was eventually brought back to the Quay, "with much apparent difficulty", where it was left until Thursday morning.

A week later the engine was still idle, as the trucks could not be obtained from the makers. "More over, the carriers (horse waggoners), jealous of the "steam horse", refused to bring in any more ore", the consequence of which was that "the chartered steamer "Mary Austen" had to leave Truro for Wales, minus her cargo of iron ore". Nevertheless, by the following Monday "the engine with one truck full of ore passed through the town on a successful trip", and deposited its load at Town Quay.

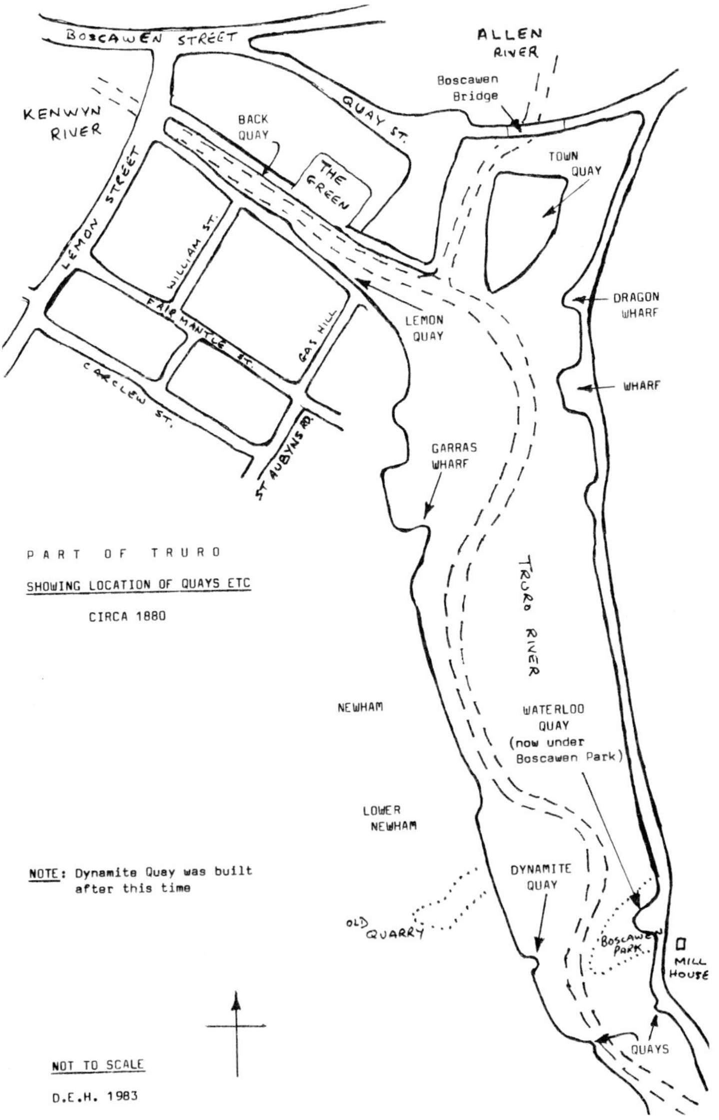

Little is known of its progress during the next few weeks but on 3rd May while taking on water at Penhallow (near Perranzabuloe), a boy named Grubb, driving a horse trap, was so engrossed with the sight of the engine that he lost control of his horse when it shied away from the engine. One of the trap's wheels went up over the hedge tipping both him and a lady onto the road. The lady received facial injuries and a broken thumb but the boy's head injuries caused grave concern. He eventually recovered but received little sympathy. Apparently the engine driver had offered to lead the horse past the engine but was refused by the boy, who,"had he then attended to his horse instead of looking at the engine, it is thought that the accident would not have taken place".

Later that week the engine is reported as constantly making trips to the iron mines for ore which was brought direct to Waterloo Quay instead of Town Quay as at first. Waterloo Quay had by then been hired by the Company from whence the ore was shipped to Swansea and other ports. The operation was, however, short lived. By 23rd May the engine had been taken off the road as a failure and the experiment abandoned. The reasons given were, "In the first place it had much too limited a power for the hills and bad roads over which it had to travel and as it was not a new machine it was very frequently breaking down". At the time the failure of this experiment was considered also as "the inadaptability of the traction engine (generally) to convey heavy loads of ore over the Cornish hills and the necessity, therefore, of tram or rail roads". A rail branch line was eventually laid to the area but the mammoth mining scheme was subsequently abandoned. As a matter of interest, the waggoners went on strike while the engine was at work, refusing to resume work until the engine was taken off the road.

What then was this engine?

Unfortunately, no reference to the engine's builder is made in any report and what details survive are scant. The engine is stated to have been of 9 nominal horse power and to have weighed 8 tons. It was capable of being pressed to 150 p.s.i. and to have cost £500, second hand. The new price was quoted as £800. It is referred to as a "traction engine" as opposed to "Road Steamer", although the phrase "accident to the fore wheel" would suggest it might have been one of the latter or possibly a four wheeled engine with fifth wheel in front for steerage.

Extensive research into the possibilities of a three or five wheeled machine revealed nothing of similar power and weight in contemporary engineering reports, although the probability of it being a Thompson Road Steamer cannot be entirely ruled out. From another view — that of an early conventional traction on four wheels — nothing is revealed to conform with known details. However, if the weight and horse power figures are changed for each other, (could these figures not have been originally misquoted?), then the possibility of the engine being an early Fowler Road Loco appears extremely likely, based on known facts. These engines were introduced in 1871 and were sometimes fitted with Greig & Avelings rubber-tyred hind wheels. Iron ore mines in other parts of the country are known to have used them. These engines cost approximately £800 and corresponds with the quoted cost of this engine.

There can be little doubt that this was probably the first traction engine (or Road Loco) used for haulage on the roads of Cornwall.

Richard Joe Grubb

Joe Grubb was born in Perranzabuloe. He farmed New Chiverton Farm, Perranwell with his brother and lived in Perranwell after his retirement. He also ran an agricultural contracting business working with various implements in and around the parish. He had a portable threshing set for a number of years. Little is known of his early work with this gear but there are people who can remember it at work in the parish.

The engine was a Marshall, number 5981, new in May 1878, of 5 H.P. with single cylinder. It usually required three or four horses to pull it. The machine was also a Marshall.

In those days the threshing contractor and farmers had a system for transporting the set from farm to farm using a combination of horses. Joe Grubb kept two heavy horses. About half an hour or so before the threshing was finished, a large whistle on the engine would be blown intermittently for several minutes. This whistle could be heard over some distance and warned the farmer, on whose farm the set was next to thresh, that it was time to bring his horses to help those with the set. It was accepted practice that the farmer next on the threshing list pulled the machine from the previous farm, while the farmer whose grain had just been threshed used his horses to follow after with the engine. Small farmers could hire horses for the purpose, if necessary. Many problems would be encountered on such occasions, the horses not being used to working with each other, or in fours, would not pull together, making the journey difficult for all concerned. Sometimes it was necessary to use six horses all together in such places as Jacks Lane, going up to Rose after threshing at Pencrennow Farm. The water would usually be drained from the engine to make it lighter for the horses and the tall stack lowered to lie over the engine while in transit. More often than not four horses were adequate for the purpose.

At this time the binder, or trusser as it came to be known, was not in use and the straw fell from the shakers of the machine down onto the ground. When sufficient straw had fallen to make a bundle about two feet in diameter, it was moved away and bound up by hand. Usually three or more men were allocated to this job. The straw was bound up by what was known as a thumb bind, made from straw, twisted into a kind of coarse rope about seven to nine feet long, and then tied around the bundle of straw which sometimes weighed nearly 50 lbs. A certain amount of skill was required in making these binds, which were then laid down for the straw to drop on.

At times some fun would be created. On one occasion at Montrose Farm, Mithian, they were threshing with four men employed making bundles in this way. Two of them, for fun, made an extra heavy bundle designed to upset the man pitching to the straw rick.

R.J. Grubb's Portable at Mithian Farm, c.1900 *Photo courtesy Mr. F. Ennor*

Presently they tossed the big bundle to the pitcher, a very big and strong man. He picked up the bundle, tossed it right over the rick into the next field and then turned round to the binders and told them to fetch it or leave it there! Yet again, on the same farm, a local character was raking dowse when he was set upon by the men binding bundles, bound up in a bundle of straw and sent rolling down the side of the sloping field.

Threshing machines of this time were not fitted with blowers to carry the dowse (cavings) clear of the machine and it was necessary to employ a man, or more often a boy, to rake this product away. This job, although relatively easy, was the most objectionable of all threshing jobs. Dust covered the operator who soon became black, but most of all he suffered from prickles which penetrated his clothes and stockings, especially when threshing barley.

Will Batten is believed to be the last man to have driven this engine, which finished working commercially about 1907 or soon after. It was then kept in a shed at Mr. Grubb's farm where for sometime it was used to cut chaff. The local boys would amuse themselves, when the engine was at work, by sneaking up and giving the regulator a touch to make the engine cuff. For a number of years it stood in the farmyard rusting away until eventually being broken up, the boiler being used as a water tank on a nearby farm. The chimney from the engine was, for some reason, taken by Mr. Grubb to a cottage in Perranwell to which he retired, and lay in the garden hedge long after his death. The machine, sometime later, found further useful life working with Warne, Grubb & Co., in which company Joe Grubb had a minor interest.

Nobel's Explosives Factory

About 1892 the national company of Nobel's Explosives took over the ailing dynamite factory of the British and Colonial Explosives Co., who operated from a complex close to Cligga Head near Perranporth. As the new company began building up production, a large volume of heavy traction engine traffic appeared on the roads linking the factory with sea and railway terminals. This traffic hauled raw materials to the factory together with coal and various other supplies for Perranporth. Dynamite and artificial fertiliser, a bi-product of the explosives production, were hauled away to Truro and other parts of the County. Frequently, sand from Perranporth was also hauled for agricultural purposes to make a return journey profitable. The nearest railhead at that time was Chacewater and the sea ports were all on the river Fal and its various branches, most particularly Devoran. To reach these ports it was necessary to haul right across the County.

More often than not, the engines were large road locomotives and the roads totally unsuitable for this class of traffic. Such roads as those were narrow and twisting with only a thin macadam surface. In most cases their maintenance came under the then Truro Rural District Council, only the main road sections being the responsibility of the County Council. During the first few years the haulage was contracted out to various firms but the company eventually purchased three large Aveling & Porter road locomotives for their own use. It is not surprising therefore, under such conditions, to find that it was not long before the roads began to disintegrate and the Authorities chase the company for recompense.

By January 1896 the roads leading from Devoran and Chacewater railway station to the Dynamite Factory were costing so much in extra maintenance that a comparative account of costs prior to, and since, the extra traffic was being kept with a view to recovering the additional amount from Nobel's. "There was sometimes four traction engines passing along the roads with 20 or 30 carts; the road was at times impassable and was in condition like a ploughed field". In those days there were no restrictions on the number of wagons pulled by an engine.

In June Nobel's were asking for compensation for damage to one of their wagons through the collapse of the bridge at Quay Road, St. Agnes, "the damage having been done by a traction engine". In actual fact the owner of a traction engine was responsible for repairs to bridges broken down by their engines. By this time Nobel's had purchased their first engine, an Aveling & Porter number 3706 (8 H.P.), a very big and heavy road loco and it was this engine that had in fact done the damage! The astute manager of the factory nevertheless had tried to place the responsibility for repair on the rural council as one of the County Magistrates had informed their surveyor of the damage and instructed him to proceed with repairs.

Before the end of the year the manager was again trying to evade

An 8 H.P. Aveling & Porter road locomotive, similar to the first engine purchased by Nobel's Explosives Factory. Although this engine did not normally work around Perranzabuloe, it does illustrate a typical road train and consequent damage that could be caused. *Photo courtesy J.H. Trounson*

responsibility to repairs at Chacewater bridge which had also been damaged by the factory's engine. In this case an engine belonging to a local contractor had passed over the bridge with its loads, just prior to the factory engine's crossing, travelling empty and rendering the bridge impassable. "The bridge was broken, the kerbing damaged and the water table crushed"! Fortunately for the Council, in this case they had a witness, together with an incriminating statement by the flagman, who had gone back to see the damage. Before the end of March 1897, this bridge was damaged yet again, "Nobel's Explosives was reported to have paid the cost of repairing .... damage having been caused by their traction engine a second time".

Not all damage was caused by the engines. Heavily loaded wagons with comparatively narrow wheels, consequently exerting a greater force on the road surface were known to damage culverts and cut into the road. In this case the manager had no option but to pay although this was done under protest.

By March 1898 the road between Three Burrows and Trevellas Downs was so badly torn up that it was estimated to cause an extra expenditure of £500 a year, "the road being in a fearful state". By today's standards £500 does not appear to be much, but by comparison to normal costs of the day, which were about £7 or £8 per mile, per annum, for normal maintenance, it represents a considerable sum, especially as the length of road was less than five miles. The figure of £500 was in fact somewhat inflated as it had included a section of County Council main road. A more accurate estimate was in the region of £250, still a lot of maintenance!

Some idea of the damage can be gained by the fact that "the road was eighteen inches lower in the centre than at the sides" and needless to say a furore erupted as to whether Nobel's were liable for the damage. Six or seven engines were quoted as working for the factory although these figures were later denied by Nobel's manager, who again cutely tried to evade liability by drawing attention to the fact that all but one of the engines belonged to contractors hired by his firm! Eventually the matter went for Counsel's opinion while the road was put in good repair.

Subsequently in the opinion of the Counsellor, "the Company was liable so far as the traffic was conducted by their order and not merely by the order of the carriers". He also advised that they were "liable for the whole period since the company entered into possession of the works and if the traffic was likely to continue, the cost of making the roads fit for such traffic could not be separated from the cost of repairing them". Although a sub-committee was formed to consider the opinion with a view to litigation the matter was shelved by the Rural District Council.

The route taken by these engines passed over Trevellas to Barkla Shop and then skirted around Mithian and up to the St. Agnes road, then on to Truro, Chacewater or Devoran as required, the latter route by way of Baldhu. By mid 1900 the upkeep of the road to Truro from the factory was still in excess of £70 per mile against an average cost of £7 per mile and early in 1901 the section of road from Perranporth to St. Agnes was described as "no better than a ploughed field, with furrows six or seven inches deep". It was in fact causing not only damage to vehicles drawn by horses but also the goods carried in them. This in spite of £134 spent on repairs the previous month on about the two miles or so of road across Trevellas Downs towards St. Agnes.

As already stated, the produce from the factory was dynamite and artificial manure. The dynamite was usually taken to Newham, Truro, where the quay from which it was despatched is still known as Dynamite Quay.* The route went right through city centre via Boscawen Street and up Lemon Street, turning into Freemantle Street and on to Newham. From there it was shipped to various parts of the world, particularly the Colonies, and used in mining operations. If ships could not navigate the Truro river to reach Newham then the longer road trip to quays with deeper draught was required. On one occasion, two boxes of dynamite fell off one of the wagons in Boscawen Street. Fortunately for Truro no calamity occurred but many timid folk in Truro dreaded the passage of dynamite en route through the City to the ship.

The fertiliser was generally sold on the home market and was usually despatched by rail from Chacewater Station. Quite naturally a small proportion of both products was sold within the County to the mining and agricultural industries.

During 1901 the Rural District Council further discussed the merits of acting on Counsel's opinion concerning damage to the roads, which were now further aggravated by the building of the

*See map on page 23 for location

Newquay railway branch line from Chacewater. Up to this time, and in spite of the fact that the rates were only about £250 per annum for the factory premises, the council had been unable to come to an amicable agreement with Nobel's concerning restitution for extra expense on road maintenance. By now the additional cost had risen to £700 per annum but the prospect of a station at Perranporth on the branch line did offer some encouragement to the council. Nevertheless, coals would still have to be hauled from Devoran. The council decided not to take any legal proceedings at present.

At the end of the year the factory took delivery of a second road locomotive by Aveling, number 4924, also eight horse power, for their own use. The two engines were further supplemented by another similar engine (number 5052), in June of the following year. Few people remember these engines working but an interesting anecdote was related to the author by one who could just remember the engines passing his home when he was a child. It shows the limits to which the engines were pushed by both their drivers and those responsible for their maintenance.

Not far from the factory, near Mithian, is a section of road in a dip, known as Barkla Shop. During the evening, after bedtime for the child, the engines could be heard returning to the factory with their loads. On climbing out of the dip the engine would gradually slow down as it took the steeper part of the hill. The driver would apply extra steam through his auxillary valve and the engine would then pick up. The child's father, a practical man, did not like this practice, often arguing that the driver was relying on the pin through the steering swivel to prevent the engine up-ending. In fact, one day, the first engine owned by the factory did just that, the securing pin having broken or come out, and the front wheels ran back under the engine which was pulling three trucks!

On the same section of road there used to be bulks of timber left in the ditches for blocking the wagons when the loads had to be split and the wagons pulled up one at a time. On one occasion a wagon ran back over the blocks and down the hill, turning into the yard of a carpenter it demolished the workshop. Fortunately no one was hurt. The ruts in the road in this area were often eighteen inches deep and the size of the engines can be gauged by the description that one of them made a showman's engine look small. Drivers were local men from the Bolingey area by the names of Tom Sanders, Nick Hubber and Jim Batten.

Another man, whose father was a steersman on these engines, told the writer that it was easier to walk than steer an engine on the roads of those days. He also spoke of the sheer size of these engines, none of which type have survived for preservation.

What were these engines like? Unfortunately, there are not any examples of similar engines in preservation. We have to rely on photographs, details in catalogues and contemporary writings for information. A good overall picture may be built up from these sources.

Built on a conventional boiler, the general appearance was of a long massive engine, not unlike the later Burrell road locos, but the looks were spoilt by the low tender.

The boiler barrel was of large diameter, quite long and fitted with three bands. The smoke box was very short with a dished door carrying the Aveling horse. The lights were carried very high. A tall narrow stack base carried a slim slightly flared stack with a rim at the top. The fire box was stayed seven rows high and the ashpan damper was fitted on the nearside. The main boiler door was cut on the nearside, in front of the firebox.

The cylinder block carried inclined top valves to the compound cylinders, which were 7″ and 11½″ diameter with a 12″ stroke, working at 160 p.s.i. The valve spindles worked in guides mounted on the trunk guides. Lubrication was by displacement lubricators mounted on the back end of the valve chests. The safety valves appear to be the same design as those on later Avelings. The regulator valve rod connected with an unequal lever, mounted on the spectacle plate, which in turn connected by another rod with the vertical driver's handle on the foot plate. A pipe on the front of the block carried live steam, through a simpling valve, to the low pressure valve chest. The exhaust steam went direct to the stack base. No cylinder drain pipes were fitted. The maker's plate with engine number, was mounted on the low pressure cylinder lagging (nearside). The crossheads ran in trunk guides and a comparatively short con-rod (2′ 9″) connected it with the crankshaft. The connecting rods passed through slots in the flanged spectacle plate. The motion was enclosed by very large plates, set quite low. A short ladder, high off the ground, led to a small platform on the offside.

The unbalanced crankshaft was carried in massive castings which also carried the bearings for the second and third shafts. These castings were fitted to the outside of the hornplates and tied together with transverse plates. All the bearings were outside the crank case and the boiler pump, which was mounted on the offside of the boiler barrel, was driven by an eccentric on the end of the crankshaft. The solid 4′ 6″ diameter flywheel was slightly dished and 6″ wide. The outside bearings allowed all the gearing from the crankshaft to the third shaft to be contained within the crank case. The two speeds were changed by a lever and forks pivoting and sliding on a casting mounted on the offside bearing casting. The sliding gears were on the second shaft. A pinion, outside the offside bearing of the third shaft, transmitted the drive to the sprung hind axle through the differential. The drive pins were square in section and the boss of the 7′ 6″ x 16″ hind wheels were cast extremely heavy. No axle caps were fitted. A band brake acted on the differential casting and was operated by a handle mounted on the tender inside the hind wheel. On the nearside, the winding drum wound the rope through a pair of very low, long guides onto the bottom of the drum. The boss of the wheel on this side was fitted for two drive pins.

As already stated, the tender was very low, without a coal guard or door. Bunker capacity was only 5 cwt. The back was slightly flared at the top. The pull on the low dragbox was transmitted from the horn plates by side boards attached below the hind axle. A toolbox was fitted on the back of the tender. A half cab, without the side boards, covered the footplate, the steps to which were on the offside with the reversing lever opposite. A very narrow belly tank was fitted. This tank was only as wide as the firebox sides and extended down to the bottom of the ashpan. Total tank capacity was 290 gallons. The steering wheel was on the offside, with seat for steersman. The steering gear and barrel was plated to the front of the belly tank and the chains wound on to a worm barrel. The other ends connected with a shallow, round, spud pan. The hitch on the cast perch bracket was hinged horizontally and heavy, domed, axle caps were fitted to the front wheels. The finish was in Aveling green, lined red and yellow, with black wheel rims.

These engines were reputed to have been very rugged but too heavy.

Aveling's claimed that they weighed 14.5-0, laden, but it is believed that they were in fact, much heavier, possibly as much as 20 tons. Their overall width was 8′ and length 19′ 3″. The stack reached up 11′ 7″. They were not the easiest of steamers. The engine must have presented a daunting sight when met by those of nervous disposition as they plied some of the narrow twisting roads on their way from Perranporth to Hayle or Newquay. In practice, they were cumbersome and a thorn in the side of the highway authorities.

The two newer engines were both similar but built less heavily and with smaller diameter (6′ 6″) hind wheels which were also wider. They had side valves which allowed large ports to be used for the cylinders (sizes of 7″ and 12″) but still working at 160 p.s.i. The steam now exhausted in the smoke box and is reputed to have improved steaming. The crossheads were still carried in trunk guides. The first three shafts were all carried in the same bracket and all the gears on the crank and second shafts were inside the bearings with the drive to the hind-axle, through the differential, on the offside but outside the hornplates. The pump drive too, was now inside the bearings.

An illustration from a work's brochure of Aveling & Porter showing an 8 H.P. road locomotive. Nobel's Explosives purchased two engines of this design in 1901 & 1902.
Courtesy Road Loco Society

Steering remained on the offside. The boiler barrel was quite long and the belly tanks were carried very low, the bottom being almost down to ashpan level. A copper top surmounted the stack but it is not certain that a half-awning, without side boards, was fitted over the footplate to which there was no door. The overall impression was a longer, lower engine and probably clumsy when in tight places. Only one engine of this type survives in preservation and that has been rebuilt as a showman's road locomotive.

The first engine was subsequently taken off the road on account of its great weight relative to the width of the hind wheels, which caused considerable damage to the roads. An old engineman who knew it described it as "a lirruping great thing".

A fourth engine, albeit a portable, was stationed at Bolenna in a shed adjacent to the well. From there it pumped a water supply from the well up to the factory, working two or three days a week. This engine has not been identified but it is believed to have been an Aveling Porter.

In the days when it was necessary to carry a red flag ahead of the engine a lucky escape is reputed to have occurred on a trip to Hayle, loaded with two wagons of artificial manure. On the west side of Camborne, at the top of the hill leading down to Roseworthy, whilst changing to a lower gear the engine started to roll and went down the hill out of gear. Only the brakes of the engine and wagons retarded the speed. The flagman attempted to stay ahead but "soon heard her coming" and moved out of the way. The train eventually reached the bottom of the hill where the engine stopped still upright but with the wagons upset.

The manager of the works was not only a hard businessman but also a hard taskmaster. One day through delays, the engine crew were unable to get back to the factory before dark. Having no lights with them, they pulled off the road at dusk and telephoned the factory to inform them of the situation. In spite of having let the fire go out, the crew were instructed by the manager to light up again and get the engine back for the morning!

In 1902 an explosion of dynamite occurred at the factory and prompted a petition signed by many influential people from Truro, to the effect that the explosives should not pass through the City's streets but after investigation the route remained unchanged and even after the railway connection had been completed it was found impractical to convey the explosives by rail as the G.W.R. company could not guarantee to get the ship to which the dynamite was being transported away on time as required by the Watch Committee.

Around 1906 the company closed the factory down but not without having made a considerable contribution toward the building of the new road from Perranporth to Truro, which at the time, was considered to be one of the best roads in the County, allowing traction engines to pass each other unhindered!

Simeon Penna

Simeon Penna was the youngest son of a family of nine children. He was born and brought up at Hendravossan Farm, Perranzabuloe. On leaving there, he moved to, and farmed, the neighbouring Treamble Farm. He was also a very skilled trapper and ran as a side-line a threshing business, with a portable engine. Short and stocky in stature, he had a mild manner, dry sense of humour and was well respected by the family.

Exactly when he started threshing is not known, nor when he purchased his own gear, but it is believed to have been well before the turn of the century. It is known that he had two sets at various times, but whether concurrently is still a mystery but seems probable. He kept a pair of heavy horses to pull the set about, often assisted by other horses provided by the farmers for whom he was threshing. All those that remember the set at work refer to the engine as being old by 1910. All his machines are known to have been Marshalls.

Whenever he started, there can be no doubt, much of his work was in Perranzabuloe and even in latter years he still threshed there, particularly around Rose and Goonhavern areas. A great deal of excitement, enjoyed here by the local children, occurred after the threshing had been completed. In order to make the engine lighter for the horses to pull, the steam and water would be blown off immediately after the engine stopped work, creating a considerable noise and vast quantities of steam. Children would congregate, waiting especially to see this event and the engine pulled away with two, three or four horses. Penna is also known to have threshed in Cubert and Newlyn parishes but, by far, most of it seems to have been done in and around the St. Columb area, even as far afield as St Eval.

Two of his brothers had moved there and must have provided a springboard from which he accumulated a considerable amount of business, in a rather unusual way. As already said, he was also a trapper and, apparently, had an arrangement with the farmers whereby he contracted to thresh their corn in exchange for the exclusive right to trap the wild rabbits on their farms. This arrangement was carried on over a number of years and in fact, Penna kept for this purpose a horse caravan in which he lived when working. It was so profitable that he is reputed to have kept a threshing set in that area on a permanent basis. In this area, especially around Whitewater, clover abounded and was also threshed with the same machines, fitted with different sieves. He seems to have employed men from Perranzabuloe as well as from the St. Columb area. Those too far from home to travel daily would sleep in the farmers' barn and kill a lamb from the farm to live on. Men coming from Perranzabuloe included Jim May, Dick Rowland and Dick Solomon. It was also in this area that he is said to have had an engine blow up. While threshing between Newquay and St. Columb, the firebox blew out of the engine, killing one man and setting the mowhay on fire. The flywheel is said to have pitched on top of the machine.

At another large farm, hear Newquay, where there was a week's threshing, Simeon Penna and two of his men, went into the farmhouse, as was customary, for breakfast. They were served a pig's head which none chose to eat. The next morning the pig's head was put on the table again and, as it was still untouched, it appeared again the following day. Simeon by then could contain his silence no longer and, as it was laid on the table by the farmer's wife, said dryly, "Good morning, I've seen you somewhere before!". The head did not appear next day.

About the beginning of the 1914 — 18 war, traction engines, which didn't require farmers' horses to pull their gear about, brought the business to a close. The engine and machine were returned to the farmyard at Treamble where they remained. The engine was housed and did further useful work by way of crushing corn and sawing timber. It finished its days rusting away on open ground. In latter years, there were two machines, one much newer than the other. About 1920 Simeon Penna suffered a stroke and didn't do much after that time. He died in 1927 aged 79. The idle machines were still there in the mid twenties and eventually rotted away. Another very old machine, parked at the bottom of Mount Hill and used as a chicken house, is thought to have once worked with the engine that burst. The newer machine is believed to have originally belonged to J. Watts of Mount.

Warne Grubb & Co.

The Warnes, like the Grubbs, were an established Perranzabuloe family. They were once heavily involved in the mining industry in Cornwall and abroad where they accumulated a considerable amount of capital. They later acquired quite a bit of farmland in Perranzabuloe parish. The Grubbs were experienced practical farmers, also owning land in the parish, and had considerable threshing experience.

John Kitto Grubb and Sydney Grubb were first cousins, their mothers being sisters. Josiah Warne of Nansmellin, near Perranporth, was also a cousin, in that John Kitto and Sydney's grandfather, John Grubb, had married Mary Jemina Warne, a sister to Josiah's father. It was, therefore, a family business in which the three cousins had a major interest and operated from Perranwell, utilising as headquarters for their operations, land and building formerly part of New Chiverton Mine, situated at the end of Call Lane. Others with minor interests in the company were R.J. Grubt H.M. Grubb and Harland Grubb. Although the Warnes had a larg

Cylinder block and motion of a Burrell "Single Crank Compound Engine"

interest in the company, the day to day management was left to the Grubbs.

In 1907 the company acquired its first engine, a Burrell, "Devonshire" type, traction, new on 28th September. This type of engine was especially adapted and built for use on the narrow hilly roads of the West Country. Although comparatively light, they were powerful little engines, rated at six nominal horse power and quite capable of driving a 4′ 6″ machine. The cylinder arrangement was a Burrell patent, whereby the compound cylinders drove a single crank by means of a large crosshead into which the two piston-rods were secured. Engines fitted with this arrangement were known as single crank compounds and were widely used in Cornwall, being noted for their economy of fuel. Examples of this type remain in preservation and can be seen in various parts of the country by those sufficiently interested. A few details at this point may well suffice others and also provide a record.

This particular engine, works number 2944, was, when new, painted maroon with maroon wheels and lined out in black, red and yellow, in the maker's own style. The smokebox door carried a ring with "The Devonshire Engine" on it. The cylinders were 5½″ and 9″ diameter with a stroke of 10″ and worked at a pressure of 180 p.s.i. The maker's number was carried on a brass plate on the valve chest cover. A pickering governor controlled the engine's speed when on the belt and on the road, the wheels were driven by the maker's double drive system of gearing, which had the differential mounted on the second shaft of the three. This method improved the wear on the last motion gearing and considerably decreased the strain on the differential gear. The hind axle was sprung which also decreased the general wear and tear. Two speeds enabled the engine to tackle the steepest hills and, in difficult ascents, the wire rope could be used to assist if necessary. When descending, speed control was assisted by a fly wheel brake and a brake acting on the inside rims of the hind wheels. The hind wheels were 5′ 8″ diameter, those on the front 3′ 7½″. The engine measured 16′ 8″ and was 7′ 1″ wide overall, and weighed about 8 tons 16 cwt laden. The company's name, Warne Grubb & Co., Perranwell, cast in brass, was mounted on a toolbox on the offside of the boiler. The engine had a threequarter length canopy when new. It was never given a name but in later life was referred to as the Little Engine. It worked with a Marshall machine on wooden wheels with a 4′ 6″ drum.

Burrell 2944, Grubb & Warne, Perranwell.
Northey's Farm, Gwarnick, Nr. Idles. c.1939 after canopy was damaged in an accident. *Photo courtesy Mrs. P. Clark*

The first man to drive this engine was Arthur Sobey, a Perranwell man. In those days, no doubt much of the work came from established connections of Joe Grubb but new business must have come in quick succession. The farmers, relieved of the problem of pulling the gear in and out of their farms with horses, would have been eager to engage the more modern machinery. Additional work could also be taken on by the company, as vital time was saved while travelling between farms. When Sobey went abroad a year or two later, Johnny Powell, also from Perranwell, took over. Soon the prospering company was looking for another set and, in fact, acquired two, secondhand at about this time (circa 1912).

One of the engines just acquired was another Burrell, similar to the Devonshire but bigger, built as a normal traction engine. This engine was purchased from J.P. Watts of Mount, who had been in

Warne Grubb & Co's Burrell 3034 threshing somewhere in Perranzabuloe parish.
Photo courtesy Mrs. P. Clark

competition with the firm. Built about a year after the Devonshire engine it was for some reason put to work by the company with an older, portable type threshing machine. The 4′ 6″ machine, a Marshall on wood wheels, is thought to have been that belonging to Joe Grubb, which had worked with his portable engine. A description of this engine is included in the notes on J.P. Watts. The other engine, a Fowler, was purchased with machine and separate binder (trusser). The engine had been supplied new to J. Hearle & Son, Agricultural Merchants, Penryn, in 1907. They are known to have had a set at about this time but it is said, that when purchased by the company, the set came from a Mr. Tripp of Philleigh, also a threshing contractor. The machine was also a Marshall on wooden wheels with a 4′ 6″ drum. Several engines of this type have been preserved; a brief description will therefore suffice.

The engine, works number 11249, was a class R, single cylinder, general purpose traction engine, new in August 1907. The 8½″ diameter cylinder with 12″ stroke was rated at seven nominal horse power and worked at 140 p.s.i. Fairly plain, with little brass and no lining, the engine was finished in dark green but nevertheless had a brass top to the stack. The maker's number was engraved on a brass plate on the side of the boiler barrel. The unsprung hind wheels were 6′ 6″ diameter, and those on the front 4′ 1½″. The motion was polished but unlike the Burrells the engine had four shafts, right

Warne Grubb & Co's Fowler 11249, threshing at Crescent Farm, Gollawater, near Perranzabuloe. *Photo courtesy Mrs. O. Stephens*

hand steering, no canopy or flywheel brake and a direct acting governor was fitted. It was 16′ 6″ long but only 6′ 10″ wide and weighed approximately 9 tons 10 cwt. Compared with the Burrells it had a slightly rakish look and was a powerful engine. It was not liked by some drivers who found it a bit light in front and, sometimes, difficult to steam when working in sheltered places, although, in the company's yard with the other engines it was usually the first in steam. It was not so much liked by the farmers, it being heavier than the Burrells in consumption of both coal and water. Like the two Burrells the engine was never given a name but was referred to as the Fowler set. As might be expected, the recently acquired Burrell was referred to as the Big set. In both cases the owner's name was painted on a piece of plate on the toolboxes. Not long after, the Fowler was fitted with a threequarter canopy and a flywheel brake, and its machine had the binder fitted.

Arthur Sobey, recently back from America, took the Big set and Will Batten from Mithian (St. Agnes), the Fowler. The company with their three sets, now could cover more than ten parishes. The Big Burrell set eventually worked as far north as St. Columb, taking in parts of Perranzabuloe, Cubert, Crantock, Newlyn, Newquay and St. Columb Minor. The Fowler set eventually took in parts of Perranzabuloe, St. Agnes, Mithian, Kea, St. Allen and Kenwyn. The job of the Little set was now to cover small farms close to Perranzabuloe parish and provide a service for the bigger farmers who needed an odd half days work done to tide them over until the main threshing was done by the other sets. This set was usually the first to come in toward the end of the season, some of its work having been done immediately after the harvest.

In 1913 Sobey left the company to set up in business on his own account and Jim May from Rejerrah took over. The 1914—18 War now hindered the company's progress, the Big Burrell being requisitioned for Army service. Jim May delivered the engine to Truro Quay from where it was taken over by Government drivers and used for baling hay. (This engine was taken in lieu of Arthur Sobey's, he having protested at his engine being requisitioned). The Government paid the company one pound a day for its use and on return it was found to be in a "brave old state", having to be completely overhauled. During the war the Little engine also worked on occasions for the war effort, pulling supplies and products to and from the Munitions Works at Perranporth. This was a short haul, being only about 1½ miles from the Great Western Railway station at Perranporth. At this time the Women's Land Army had been formed and one day, the Little engine working at Causilgey Farm, Tregavethan, was used for a film depicting the women at work. When the film was made, normal threshing was stopped and women changed places with most of the men at their tasks. The whole thing was filmed with only the first and second men, all the other "workers" being women. As soon as the filming was finished the men resumed their places. Later, the film was shown in a Truro cinema, "a lot of guff" in the words of one of the men who was temporarily relieved of his task. The engine was threshing at Reen Manor Farm on Armistice Day. When word came through that the war was over, the set was closed down and the men left work for the day.

After the war business became more normal and the Company flourished. Johnny Powell was still driving the Little engine, Jim May the Big Burrell and Will Batten the Fowler. In 1921, under new legislation, the engines received from Cornwall County Council the registration numbers AF3579, AF3580 and AF3581 respectively.

At this point an insight into the work and conditions of the job in the early twenties may be appropriate.

It was customary for the first, second and, sometimes, the third men to be employed by the Company and the other workers, known as "followers" (sometimes "slingers"), were paid by the farmer. The Company's men signed to work the full season and, as a consequence, received regular pay irrespective of the amount or conditions of their work. The followers were taken on daily, usually on the first man's advice to the farmer, he having worked with most of them and knew their capabilities. The followers were a tough lot and would wait outside the farm gate each morning, looking for work. Some were quite elderly and the farmer or first man would pick out those wanted for that day, usually six to eight in all. They would expect to earn from four to six shillings (20 to 30p) for their day's work. If they were lucky, the farmer would sometimes offer a few the job of thatching the ricks, immediately after threshing. Ricks of baled straw were also thatched. Good workers could expect to follow the machine over about a six month period. During the day if the weather changed for the worse and threshing had to be stopped, these men would only be paid for the hours worked before stopping. On such days, the Company's men would perform maintenance jobs, under the tarpaulins. Sometimes they would leave work early but "the big farmers would make sure they were there all day". It should be remembered that all these men worked on what was basically a winter job and were out in all winds and weathers, often miles from home, their only transport being a bicycle or shank's pony!

Relating some of his memories to the writer, an old follower said, "It was all slog in those days". They usually rode to work on bicycles "cycling four abreast, chatting and laughing. If the wind was blowing from behind the engine, they would be black as tinkers by the end of the day". As a boy he remembered "thumb binds weighing half a hundredweight or more in the ricks". He remembered two old timers who followed the same machine as himself. Both were hard of hearing, one always dressed with a hessian bag in front, tied at the waist and another over his head and shoulders, with a bowler hat on top. This man usually cut the binds and would shout from the machine, "I can take them in as fast as you can throw them up". The other man would, at the same time, be shouting back, "I can throw them up as fast as you can take them in". The shouting match would be kept up all day. Another, reminiscing about the same pair, related how one day when threshing at the top of Kenwyn, near Truro, these two characters disappeared in the dinner hour. When work was about to resume they reappeared, having walked down to the City Inn, Truro, to have a few pints. Apparently this pair, known as Joe Rhubarb and Old Man Hosking, used to walk everywhere to their work, usually from Penwartha Coombe near Perranporth.

A description of the general tasks that followers did is perhaps also apt. Generally, there would be two men pitching sheaves from the corn rick to the machine, where the binds were cut by another before being passed to the second man who fed the machine. After the sheaves had been threshed and the straw bundled by the trusser, another follower pitched the bundles to the straw rick, built by yet another follower, who usually specialised in this job. At the same time there would be others looking after and carrying the corn sacks. In addition, the sacks had to be transported to the barn and water brought to the engine. On some farms the straw was baled and ricks built with them. When National Insurance was first started, the farmers were reluctant to thresh on a Monday as, due to this system of employment by the day, the farmer who employed the man on that day of the week was responsible for paying the whole of the employer's part of the stamp.

On the threshing set, the first man's duties were to look after the engine and get the set from farm to farm on time. He was in fact in charge. The second man looked after the machine, feeding it during threshing and generally helping the first man. The third man looked after and tended the binder (trusser). The first man would occasionally relieve the other two for a while during the day, to give them a break when they would at the same time keep an eye on the engine. Sometimes the third man was paid by the farmer. Naturally, the first man was in close liaison with the farmers but there were frequent rows between them, the bigger farmers sometimes taking a high-handed "holier than thou" attitude. Some would not even provide soap for washing before going home. Others would try to outdo the coal arrangement.

It was also customary for the engine to leave the farm with the bunker full of coal, the farmers being responsible for providing coal when travelling and threshing. Once after finishing threshing on a

Saturday afternoon one of the drivers decided to wait until Monday morning before moving the set to the next farm which was not far away. As usual, he filled his bunker with coal, agreeing with the farmer that the set could stay there over the weekend. On Monday morning, the second man lit the fire and later drew attention to the fact that the bunker was only partly full when he had arrived earlier that morning. The first man went down to the cowhouse to see the farmer about the matter. On being asked, "What happened to the coal?" the farmer looked rather guilty and tried being evasive. The discussion was pursued further and eventually, the farmer told him, "The coal was in a bag under the dowse pile". Another farmer had the set left on his farm all night because he would not provide a bag of coal to properly fill the bunker.

When Johnny Powell left the Company, Jack Juleff from Gorran took the Little set for a season. Baden Powell from Perranwell next drove this engine for three seasons and Dick Rowland from Hendra Croft took it after him. In the mid twenties John Rowland, a brother of Dick's, succeeded him and later it was worked by Preston Thomas from Cocks Hill for a number of years. The last man to take the set was Will Annear from Perranwell. He drove it for the last two seasons that it worked with the Company. On the advice of the boiler inspector, he refused to take the engine out anymore, the flange on the boiler being completely corroded away around the bottom half of the joint with the smoke box, which was also very weak. The engine then lay in the yard for a year or so.

Generally this engine was liked by its drivers but, if the wind was in front of the engine, it was often necessary to put a piece of galvanise up to stop the smoke blowing back under the canopy. The original stack had been replaced after the first few years with the maker's cast iron type which was slightly short of the canopy. A day's threshing could be done on five hundredweight of coal. Its working life was not without the hazards that beset tractions working on country roads in hilly districts. On one occasion, going towards Newquay, this engine slipped away at Rejerrah (the old road), and finished up looking up the hill with the machine looking down. On another, at Liskey Hill in Perranporth, the driver had got up over the steeper part and screwed down the brake to change gear. When the engine was out of gear, the set started slipping backwards and the machine slewed across the road with the engine against it. One of the wheels of the machine got crushed. The screwed bush on the brake rod had stripped its threads, causing the ensuing accident.

The most serious accident occurred at Bussavean Farm, Shortlanesend, in Kenwyn parish. John Rowland an old but experienced driver was pulling the machine into the farm, approached by a lane on the side of the steep valley, near the bottom of which was a sharp corner. He had encountered difficulty on the corner and decided to use the rope. Moving further up the lane he reversed the engine toward the side of the valley. Whether he did in fact trig the wheels is not known but in the event, the engine either came over the trigs or slipped away from the side of the valley when the rope was under strain. The engine ran backwards down the side of the valley and turned upside down into a wide ditch. Relatively little damage was done to the engine as it had its fall broken by thick mud in the ditch. The canopy was smashed, the boiler lagging crushed and a stub of the front axle broken off. Rowland was even luckier. He was still in the engine when it came to rest but apart from severe bruising he was unhurt. Nevertheless, he had to be pulled out of the engine which was slowly sinking in the mud and would have been trapped within ten more minutes. The engine was righted by the Big Burrell and a temporary front axle fitted, enabling it to be towed back to the yard. It worked for a while without the canopy which was subsequently rebuilt and refitted.

Most of the work that this engine ever did was threshing but it was used sometime around 1930 to pull the Cornish boilers out of Wheal Vlow Mine on Gear Sands. Sometimes it would be hired to cut chaff for farmers, "cuffing all the time" on this hard work. Once when working near Trevemper in Crantock, a wing valve within the injector clack box broke off, allowing the steam and water in the boiler to be blown back through the water tank. The fire was pulled, the valve removed, brazed and refitted, and the engine filled with water by hand in readiness for next morning. The driver and his second man arrived home at one o'clock in the morning. Sometime during the thirties the engine was fitted with a new firebox by Argalls of Truro, who also did any general heavy engineering jobs when required. Towards the end of this decade work started to become scarce, the internal combustion tractors were beginning to take over and the available work was not really sufficient to keep the crew happy. The set was now out only about 60 days a year and by the early forties it was idle as already said. Nevertheless, it was not sold until the other sets were too.

After the first War, Jim May, as first man, had the Big Burrell back. He was succeeded in the early twenties by Dick Eley of Perranwell, who drove it for a few seasons, and he in turn was succeeded by Paul Clark from Perranwell who drove this engine for the rest of the time that it was with the Company. Eley had lost an eye when working for another contractor. Leaning out over the side of the engine to see better while moving the set in the dark, a dog-rose from the hedgerow caught in the flywheel and whipped back into his face, destroying one of his eyes.

Burrell No. 3034 at Grubb & Warne's yard, Perranzabuloe, 24th August, 1938.

This engine has been described as a lovely engine by men that worked with it. It also frequently drove a baler with the machine, especially in the thirties, when the Company purchased one of their own, manufactured by Hornsby. Around this time it frequently worked at Treamble Mill, out of season, driving a crusher and other machinery. Like the Little engine, the original stack was replaced by one of cast iron and the firebox was renewed in the thirties.

Argalls also did this job and used roof stays to the boiler shell instead of girder stays as originally fitted. They also restraked the hind wheels at the same time. The engine did a lot of work being out for full seasons, but was never involved in any serious accidents. Like the Little engine, it had slipped away on Rejerrah at times but without serious consequences. The crew eventually resorted to using frost spikes in the nearside hind wheel and fish plate bolts (from Great Western Railway fish plates) in the other on this hill, when they judged it necessary. About 1938, a motor cyclist mistakenly tried one evening to ride between the machine and baler. Because the baler used to foul the trusser of the machine when going down hill or turning sharply, if towed behind the machine, it was always pulled in front. The complete set was in the process of turning into the yard at Perranwell, when the accident happened. The baler had no lights on it at the time and the motor cyclist, a Mr. Kempthorne, was in hospital for several months. This engine was the most active at the time the Company sold out.

Most of the men who had driven the Burrells didn't like the Fowler, but those who hadn't loved it. Will Batten drove this engine until about 1932, when Jack Juleff, now moved to Trevellas, St. Agnes, took over. He remained with the set until a considerable decline in the amount of work the Company did had taken place, finishing suddenly after an argument about which engine should be working most. Will Annear then took over when necessary for the remaining year or two before the set was sold with the others. Juleff always regretted not purchasing the engine for himself.

This engine at times also baled straw when threshing, the baler being hired from Carlyons of Mount Hawke. It once astounded all that witnessed the event, by pulling both the baler and machine up across a field near Tregavethan, (part of the Kenwyn valley) without using the wire rope, which contrary to everyone's expectations was unnecessary. It too, was reboxed in the thirties at about the same time as the others, all three in fact being done within a period of about twelve months. The new box for the Fowler cost £150 — fitted! Like the others it had its share of mishaps. Coming down from Mithian school towards Barkla Shop, St. Agnes, the engine ran away and a nasty accident was averted by the steersman's presence of mind in turning the set off the road towards Mithian village. No damage was caused to the engine or machine which had lifted its inside wheels coming round the corner of the road junction.

Another time, coming up Four Burrow Hill, the engine slipped back, jack-knifing the machine, the canopy catching the trusser and breaking three cast iron gear wheels. Another day the turntable and perch bracket separated, "putting the engine down on its knees". Yet again, one of the hind wheels broke into a culvert when turning into a field at Anchor Farm, St. Agnes. The hole was so big that the engine bridged it and had to be pulled out by another, the Big Burrell.

Such mishaps to engines were part and parcel of the work that the men that handled them had chosen. Respect for them still lives on even though they are dead and gone. These characters differed but their tenacity was the same. There was Jim May, a quiet, patient, good living man, always ready to show or tell a youngster how to do a job. Paul Clark, a name synonymous with that of the Company, probably having spent more time with it than anyone else. Jack Juleff, a masterman with his engine, especially in tight places, but hot-tempered and impatient. They worked with tractions nearly all their lives. Other younger men who worked with these men and handled the same engines, now find people of their grandchildren's generation cannot believe what they had to do to earn a living, or some of the conditions they had to tolerate. The fiery Jack Juleff asked if he could wash his hands before having something to eat, after bringing the set into a farm near Mithian one evening. The rather proud farmer's wife refused to allow him to do so. He got back on the engine and told the farmer he was going out again. The farmer shut the gate and protested but was told if the gate wasn't opened he would drive through it. The set went out and on to the next farm, the farmer having opened the gate again! At another farm, another driver had seen the body of a mouse drop into a teacup from the jug from which tea was being poured. At another farm the crew were expected to eat in a room with a sick pig under the table. When working some distance away from home it was not unusual for the men to leave as early as four in the morning, cycling to their work place and getting the engine in steam, ready to start work at eight. Sometimes accommodation was offered but was not always clean — one man slept on the machine one night in preference to the bed he had been provided with. At such times, when far from home, their bicycles would be carried on the needles of the trusser if the set was travelling. In spite of their hardships these men in later life look back to those times as good old days.

Bad times there may have been but on the other hand, is not job satisfaction proportional to the effort put in?

At the end of each season the sets were always brought back to the yard where the machines were cleaned out and, later, checked by fitters from Marshalls who built them originally. Even so, it was not unusual for a bearing to catch fire in the season. Strangely, the engines didn't receive any such expert treatment, the work not being within their drivers' capabilities being carried out by Argalls as already said. All the machines had trussers fitted by the start of the first War.

Fowler No. 11249 and Burrell No. 3034 at Grubb & Warne's yard, Perranzabuloe, 24th August, 1938

The day-to-day management of the Company was delegated to John Kitto Grubb. A bachelor with a gruff voice, he walked with a slight stoop and was noted for his parsimonious attitude to money. Nevertheless, he was found to be fair. At the end of each week he would pay his men a few shillings "to go on with" and, at the end of the season, the first man would hand in his record book which would be checked as to what balance was owed and the matter squared up.

He travelled about the countryside on a bicycle taking orders and money for threshing. Generally, he was also reluctant to spend money on the gear, never spending more than was absolutely necessary. As time went on, he gradually bought out all the other shareholders and in latter years the firm's name, stencilled on the tarpaulin covers of the machines, was changed to Grubb & Warne. By the early forties he was in his seventies, the gear was getting old and business even slacker. A sale in the Company's yard in 1940 concluded its work.

At the sale the Little engine brought £200 and is reputed to have been bought by a concern near Liskeard. Eventually, however, it found its way to Norfolk in the ownership of E. Barbee, Alby, Erpingham and in whose name it was last registered, the licence expiring in December 1948. It was then disposed of to L. H. Witham, Holt Road, Cromer, where presumably it ended its days.

The Big Burrell was purchased by its first and second men, Paul Clark and Jim Nicholls of Goonhavern for around £300. The machine and baler brought the total lay out to around £700. They then commenced in the threshing business on their own account, a short record of which is included later in these notes.

The Fowler was acquired by E. Mewton of Mount Hawke, St. Agnes, bringing approximately £400. He used it for haulage purposes until the end of 1946, after which time it was cut up for scrap.

J.P. Watts

John Watts was born in Perranzabuloe and lived at the bottom of Mount Hill, near the parish boundary with Cubert. Two of his brothers-in-law had threshing concerns, Simeon Penna of Treamble, Perranzabuloe and James Penna, Trelowarren, near Helston. These outfits were run as sidelines to their owner's normal business. John Watts was a mining man and had also done some water divining on occasions. One morning in September 1908, he told his wife that he was going over to see James Penna. He returned later that day with a new traction engine.

The engine, a Burrell, was purchased by him through the agricultural merchants, Dowrick of Bodmin, who were also local agents for the company. It was of six nominal horse power, with cylinders on Burrell's patent single crank compound design, and was allocated the works number 3034, despatched on the 11th September and costing £580. It is believed that the Marshall machine he purchased was also new at that time. Several people remember the maroon engine, lined out in Burrell's noted style, being the talk of the place for a while, it being thought to be "a big thing for an engine to come new in those days". The set was kept in a meadow below Watts' cottage and he, a meticulous man, forbade anyone to touch it. His children also were not allowed anywhere near it. Such was the pride of enginemen in those days.

When ordering his engine no doubt Watts was influenced by his brother-in-law James Penna, who already had two Burrells at work in other parts of the County. Many of the specification details are similar. Bigger than the Devonshire type, this engine was pleasingly proportioned. The six foot diameter hind wheels were slightly larger than usual for this type of engine, although the width was the same at 16″. The fore wheels were standard 3′ 6″ diameter with 5″ x ¾″ steel tyres and the axle centres were set 9′ 7″ apart. The 6″ x 10″ cylinders, with 10″ stroke, worked at 180 p.s.i. and made the engine also more powerful than the Devonshire type. A Pickering governor controlled belt speeds. Boiler layout was fairly standard and had a grate area of 4.6 square feet with 38 tubes 1½″ diameter x 5′ long. The injector clack was fitted to the firebox side and that for the pump halfway along the boiler barrel, but above the centre line. A flywheel brake was fitted together with a transmission brake acting with the differential on the second shaft, and activated from inside the offside hind wheel. The hind axle was unsprung and geared for double drive to the wheels. The engine carried a threequarter length canopy, brass bands trimmed the cylinder lagging and a copper top the stack. A wooden box was mounted on the back of the tender; a similar box fitted on the side of the boiler carried a cast brass plate bearing the name J.P. Watts, Perranporth. As fitted, the engine weighed 9 tons 16 cwt. (laden), was 16′ 9″ long, 7′ 7″ across the hind wheels and the top of the stack 11′ 1″ high. The engine never had a name.

It is believed that John Watts had learnt to thresh with Simeon Penna, with whom he had done a lot of threshing and was now competing for work. Business took in farms in Perranzabuloe, Cubert, Crantock and the Newquay area, although most of his work was in the two former parishes. He drove his own engine and employed a local, Edwin Penna, as second man. A man from Rose by the name of Edward Trevethan also worked with him.

He was known to be an exacting man, always very punctual, and early one morning, an engineman, whose gear was threshing on the other side of a valley, decided to be in steam before him. Steam was successfully raised and the engineman barracked Watts by continually blowing his whistle, until he also had started threshing. Because of John Watts' nature, it created quite a bit of fun with the followers on that day.

Watts' enterprise did not bring him a rich reward. Competition was great and a price cutting war was carried on, especially with the neighbouring concern of Warne Grubb & Co. After four seasons, disillusioned by the price cutting, he called it a day, selling his engine to Warne Grubb & Co. in 1912 or early 1913. It is believed the machine was later purchased by Simeon Penna, whose gear was by then "getting all runned out".

It has been said that John Watts was in partnership with S. Penna but there is no evidence to confirm this statement. The receipt for the purchase of the engine was made out to J.P. Watts and no one else. An old threshing machine at the bottom of Mount, which for sometime housed chickens, is believed to have been once owned by S. Penna. Afterwards, he went abroad mining and on return moved to the Rock area, returning again to Rose for a while before his death at Rock.

The engine, after about thirty years with Warne Grubb & Co., was

Burrell 3034 with ex Warne Grubb & Co's drivers, the late B. Powell (on front plate), and P. Thomas with W. Annear standing. The engine is preserved in south Devon and is now named "Pride of the South Hams". *Photo the Author*

sold to Messrs. Clark & Nicholls of Goonhavern, Perranzabuloe in 1940. About 1948 it was purchased at scrap price (£8), by M. Roberts of Carland Cross, near Mitchell. Although he never used the engine, it can be said that it was saved from the scrapman at this time. Later, in 1950, the engine was sold to Messrs. Lello Bros., Hayle, who used the boiler for soil sterilisation until 1956, by which time the preservation era was established. It was then purchased by Dr. Dyke of Rilla Mill, Callington. He sold it around 1959 to Mr. B.D. Camp of Modbury, Devonshire, who has done a considerable amount of restoration work (including fitting a new firebox), to this fine old engine. On occasions it has been used for threshing on his farm and the engine now bears the name "Pride of the South Hams".

Arthur Sobey

Arthur Sobey was born in Perranzabuloe. Little is known of his early life but it is believed that he worked with the portable engine and machine of J. Grubb, Perranwell. Certainly in 1907, when the firm of Warne Grubb & Co. purchased their new Burrell traction engine, he was the first man to drive and work this threshing set. He then went to America for a while where he accumulated some capital, and on returning rejoined the Company. Whether his enterprise in crossing the Atlantic was inspired by a desire to run his own concern is speculation, but it is also certain that he took a fancy to the larger Burrell since acquired by the firm and now entrusted to his care. He drove this engine for a while and was so impressed by it that it was not long before he was taking measurements and building up a specification for an engine of his own, soon placing an order with Messrs. Burrell & Sons, Thetford. He had also given thought to the technicalities of threshing and decided to use a Clayton & Shuttleworth machine. Exactly why he chose this machine is not clear, but some say he was particularly interested in wheat and thought it cleaned the grain better than the more popular Marshall. This was to be the only machine in the area that was not a Marshall. At that time he was living at Lakewell Cottage, Perranwell, from where he operated until retirement.

On 29th July 1913, traction engine number 3500 was despatched by rail from Burrell's works, destined for Perranporth Station. Whether the machine was also despatched from Lincoln at about this time is not certain, but more than likely. In due course the goods arrived.

The engine was of six nominal horse power, built as a single crank compound with cylinders 6″ x 10″ and a stroke of 10″ working at 180 p.s.i. The cover on the valve chests carried the maker's name and number on a brass plate. The motion was polished bright and the cylinder covers were nickel plated. Brass strips trimmed the edges of the cylinder lagging. A flywheel brake was fitted and, to Sobey's specification, brakes acting on the rim of the hind wheels were operated by a very high hand wheel on the offside of the tender. The hind axle was sprung and driven by Burrell's double drive system. A baffle plate hung inside the firehole door and a brass top crowned the stack. An extra half width damper was incorporated in the rear end of the ashpan, operated from the footplate by a chain. The bottom gauge cocks were fitted with ball valves. The clack boxes were set high up on each side of the boiler barrel. The engine was otherwise much the same as the bigger Burrell of Warne Grubb & Co. (not the Devonshire type), but nevertheless created the impression of being bigger and was, in fact, one ton heavier at 10 tons 16 cwt. A threequarter length canopy, without safety valve pipes or lettering, protected the engine and driver, the whole was finished in maroon with red wheels (6′ hind), and black hub caps, all lined out in traditional style. The owner's brass name plate fastened to the toolbox read Arthur Sobey, Perranwell, Perranzabuloe. In years to come the stack was to be replaced by one of the cast iron variety.

The machine was on wooden wheels and fitted with a Ruston Hornsby trusser.

No doubt Sobey, by his own reputation, drew some farmers away from Warne Grubb & Co. when he started, and he succeeded in building up a substantial round which kept him busy for the full season. His round included farms in Perranzabuloe, Cubert and Crantock parishes. Occasionally, the odd farm was threshed in Newlyn East and in latter years he travelled as far away as the other side of Newquay. In later years most of the threshing was corn but just under a third of the time was spent on wheat, grown by some farmers for the straw, with which they thatched their hay and straw ricks. In practice, the machine was prone to winding and the shakers were too short. Sobey had modified these with some success, by increasing their length, but never completely overcame the winding problem.

The engine was not used for anything else but threshing except on occasions if the farmer insisted on baling his straw. Then the baler belonging to the Newlyn East Co. was borrowed and worked from the machine, much to Sobey's displeasure. In January 1921 the engine acquired the registration number AF 3341. In spite of his affection for the engine, it never had a name.

Sobey's nature dictated that he had to be at the centre of things and from the start he drove his own engine, taking a younger man, Jim Westlake, a Plymothian who settled in Perranzabuloe, as second man. During World War 1, Westlake went for service, and a man from Rejerrah called Alf Pope moved into his place. At this time many engines were requisitioned for army work. Apparently

A. Sobey's Burrell 3500 at Lambourne Farm, Perranzabuloe.
Photo courtesy Miss. D. Westlake

the authorities had planned to requisition Sobey's engine, but instead used one from the neighbouring firm of Warne Grubb & Co. In the opinion of the Government Inspector, whose job was to ensure that all engines were kept in sound condition for emergency service if necessary, Sobey's gear was the best kept set in Cornwall.

After the war Westlake returned to his former job as second man. By then Preston Thomas from Cox was third man, later, Will Hopper from St. Erme, by then living at Perranwell, worked as third man and was eventually succeeded by Gordon Tonkin of Rose. He later succeeded Westlake who had taken on a farm and during the last two or three seasons, before Sobey sold out, he worked as first man with Edward Hosking of Goonhavern second. By this time their employer was suffering badly from chest trouble and unable to work. Tonkin was the only employee to have ever taken overall charge of the set. He was considered a cautious driver, but in spite of this Sobey could not, or would not, rest happily and was continually worrying about his gear.

Undoubtedly, he kept it in immaculate condition and his reputation for doing so lives on more than fifty years later, in the memories of all who knew him. He was considered to be a good engineer and was always very careful with his gear, even to the extent that it sometimes got him into situations which, had he been less cautious, would not have arisen. In the words of a very experienced threshing driver, "It was a nice engine but he was afraid to work her". He wouldn't allow anyone on the engine (except in latter years G. Tonkin), and "everything had to be 'xactly for him". One old engineman remembers the set as "lovely gear" and the engine "like a clock". His only driver said, "It was a handsome engine, so quiet when working that the only sound to be heard was the ticking of the mechanical lubricator", and this was after more than ten years of work.

The engine and machine were kept side by side in a shed, adjacent to the lower side of Sobey's house. At the end of the season the gear was put away and virtually no one was allowed in the shed. Immediately before shedding the engine the tank was filled and the boiler steamed. The injector delivery pipe was disconnected at the clack box and a piece of hose connected to it. By judicious use of steam and water the whole engine was steam cleaned. This operation was carried out every year and was obviously condusive to the respect he gained concerning the look of his set. In the words of an old farmer, "He kept his set so clean that it was like new, and you wouldn't know it had been threshing." Another described it as "marvellously kept gear". A follower, at different times of several machines, said "there was some difference in his and other gears". After steam cleaning the engine was put in the shed with the machine and the firebox lagged in black oil. An oil laden rag was pulled through the tubes. Any necessary work was carried out and the set was not brought out again until the day before the start of the next season's work.

Sobey has been described as a "smart man" and "a good man for his job". He taught a lot of men to thresh, especially in the early days. "Everything had to be right for him" and this gained him a reputation of being "awful fussy" by some men, but those who were closest to him found the man "comfortable to work for". He was somewhat close and generally regarded as "a skinflint about money" by those who worked for him as well as those for whom he worked. On one occasion, fearing the farmer might question his bill for a day's work, he was known to have gone up onto the machine and said to Westlake, "Why are you throwing the sheaves in so fast? If you don't slow down we shall be finished before five!"

Around the 1923/4 season, farming generally started to fall away, Sobey was already suffering badly with his chest and over sixty years old. He wanted to get out and offered his set for sale at £550 to more than one person he thought capable of purchasing and running it. Toward the end of succeeding seasons he would try to persuade others to take on the business. Soon after he became unable to work, business was falling away and around 1934 he finally found a buyer. The purchaser who had agreed to buy it unfortunately fell ill, and Sobey had offered to release him from the agreement, but nevertheless the deal was completed on his recovery. Sobey informed his driver one evening at ten o'clock that "a new man would be coming tomorrow to take over the engine". When asked why, by the now jobless driver, he was told simply "the gear was sold to a man up the County." At that time the engine was at Lower Lidra Farm, Rejerrah, ready to thresh next morning. The purchase price is reputed to have been £500. By then, it was the only set in the parish competing with his ex-employer, who was still in business.

When the engine and machine finally pulled away on its way to the Bodmin area, Sobey is said to have walked up the road behind it with tears coming down his face. "There was no engine like a Burrell for Sobey."

Note: The set was purchased by James Penno of Ruthern Bridge, Bodmin. He sold it to F.H. Harris of West Taphouse, Liskeard, circa 1938 and still later, in 1948/9 it was purchased by Vick Bros. Metals Ltd., Stonehouse Quay, Plymouth. The last road licence expired in December 1949, and regrettably, Burrell 3500 was broken up for scrap on 8th May 1950. Sobey remained at Perranwell and his only child, a daughter, in frail health, died soon after him.

The Munitions Factory

Between 1914—18, the old Dynamite Factory premises at Perranporth were used for the manufacture of explosive articles for use in the war. Over a thousand women and about fifty men and boys were employed there, many of whom travelled by rail from as far away as Redruth and Truro. Most of the supplies were brought by rail to Perranporth Station from where they were transported with the aid of traction engines to the factory. Similarly, most of the products were transported to, and despatched from, the station. Two or three old engines were requisitioned for this purpose and were driven by local drivers. Other local engines occasionally worked for the odd day or two as necessary.

One of the requisitioned engines was a big Aveling and Porter, which had been kept at the factory, in the higher boiler house, since Nobels finished producing dynamite. It was rusted badly and in a rough state. Dick Eley, who came from a mechanically minded family and was an experienced driver, undertook to get the engine on the road again. This accomplished, he drove the engine which is reputed to have never let them down during the war period. It is believed that this was the second of the three engines formerly owned by Nobels, number 4924 new in 1901 and of 8 H.P.

Another engine, a 6 H.P. Burrell, number 2464 came from somewhere away from Perranporth. It was believed to have been requisitioned from R. Dumma & Sons, Jedburgh, and was new in 1902. Although not as old as the Aveling it frequently broke down going up St. George's Hill. Jim Batten from Perranporth drove this engine.

The route that these engines took (usually pulling only one truck at a time) was from the Railway goods shed just off Station Road, along Station Road turning left at the junction with St. Piran's Road, travelling through the porth and then up St. George's Hill to the factory. The roads were in those days macadam and with such continuous use they soon became very deeply rutted, even to a depth of nearly eighteen inches in places. The engines were continually very dirty especially in the winter. Indeed, in the winter months it was necessary to employ men with scrapers to clear away mud and slime for the engines. At these times St. George's Hill was also so deep in slimy mud that it was impossible to walk up the hill, a detour via Bollenna Lane being necessary if walking to the factory. Many a young boy lost his boots in the mud of the roads which were said to be like ploughed fields. In the summer, clouds of dust were created as the engines plied to and fro with their loads.

Other engines that worked at various times were the Little Burrell of Messrs. Warne Grubb & Co., Perranwell and a big single crank compound traction engine owned by Messrs. Simmons Hodge. This engine was driven by a tough character called Jack King, a rather reckless but nevertheless competent driver whose exploits with his engine have become legend. He frequently brought supplies from further afield. (This engine was No. 2897).

During this period Dick Eley was subjected to quite a bit of leg pulling for awhile. Apparently he had complained to the factory manager about his engine, the conversation going thus, "The engun id'n the best of pullers and another thing he won't stop." "Well, keep'n goin' then," was the reply he got.

Towards the end of the War these engines and trucks were replaced by petrol lorries.

F. Barrett

F. Barrett was a mining engineer who came from the Calstock area about 1916 and settled in Perranporth. He owned an old traction engine which was purchased for hauling mining machinery about Cornwall.

This engine was new in December 1887 and was a Fowler, class B1, road locomotive of 8 N.H.P., works number 5483 and named "Gamma" by its first owner, The West of England Granite Co., Penryn. As a matter of interest, this was the second recorded road locomotive granted a licence by Cornwall County Council. While in the ownership of Freeman Quarries (who took over the former concern), this engine is reputed to have pulled an engine which had broken down, plus its load, a Cornish engine bob, to its destination, a gross load of over 40 tons exclusive of its own weight.

Still later, it passed into the hands of a celebrated old engine contractor, Charlie Carlyon of Greenbottom in Kenwyn Parish. He performed many haulage feats with the engine, once hauling a portable engine behind its normal load of three wagons full of red bricks over the steep Cornish hills! Carlyon is said to have regularly held back this engine and its loads when going down hill, with one foot on the backhead of the boiler and both hands pulling on the reversing lever, in addition to the brakes fitted to the engine and wagons.

F. Barrett used the engine for about three years, mainly hauling, installing and salvaging mining machinery. It was also used on salvage work on a ship wrecked at Holywell Bay.

Two wagons were used when necessary, one of which could be converted to a boiler carriage. On salvage work at Bolingey Mine, the sheerlegs of the Cornish engine were taken down (using the wire rope), in such a manner that they could be wholly used again.

This engine, as already stated, was a class B1 road locomotive, one of the forerunners of the successful "B" range of engines produced by Fowlers, and which eventually developed into the finest road locos ever built. It is interesting to note briefly some of the details. The engine had a narrow high up appearance, the 7′ 0″ hind wheels being noticeably narrow (16″ or less). It was originally fitted with a powerful flywheel brake and also a transmission brake acting on the third shaft. It had two speeds and was not sprung.

There were no guides for the valve spindles and the weigh shaft was mounted directly onto the cylinder block. The governor was a large bulbous affair mounted centrally over the motion. A single hole draught hitch was rivetted to the tender and the boiler lagging ended behind the rivets of the front tube plate. The reversing lever was on the nearside.

Fowler 5483 lying in Warne Grubb & Co's yard, once owned by F. Barrett.
Photo courtesy J.H. Trounson

About 1919, the engine, together with the two wagons, was steamed under orders of the road authorities, from the old coalyard at Perranporth up to Messrs. Grubb & Warne's yard at Perranwell, where it lay rusting away until about 1932 when it was cut up for scrap. Wilfred Barrett drove the engine on her last journey by which time, after a hard life, it was worn out.

Clark & Nicholls

Paul Clark had worked with Warne Grubb & Co., since he was a boy (service in the first War excepted). Jim Nicholls was younger and had worked as his second man. They both came from near Rose and at the sale of Grubb & Warne's gear they purchased the engine, machine and baler with which they had worked. The engine was the "Big" Burrell number 3034, and was again based in the now rented yard of the old firm.

Burrell 3034 at Rosehill Farm, Rose, c.1943. Clarke & Nicholls, Goonhavern.
Photo courtesy Mr. J. Nicholls

The partnership was now the only contractor in the parish still using steam. Business came from Bolingey and Gollawater within the parish and also farms in the parishes of Cubert, Crantock and St. Columb, together with a few places around Zelah and Shortlanesend in St. Allen and Kenwyn respectively.

In 1941 Will Annear, also an employee of the older firm, joined as third man, and about 1943 the concern was offered a new Massey Harris tractor with Allen machine from the War Agricultural Committee. A driver for the tractor was also provided. With the acceptance of the offer this situation allowed the firm to take on more work. Paul Clark went out with the new equipment, while Jim Nicholls took care of the steam set with Will Annear driving the engine. This arrangement was to continue all the while the engine was still in use.

During these last few years of the engine's commercial life, its work was still occasionally spiced with the hazards that beset such enterprises. Paul Clark had had the engine slip away coming down Church Hill, Perranzabuloe with the machine, and ended up against the side of the hill. This route was usually avoided by taking the longer way down over Cox Hill and up through Reen Cross to Goonhavern. The strakes on the hind wheels were well worn by this time, to the extent that only part of the width of the wheel was in contact with the road at times. The engine once slipped away on a short dip, known as Downs Hill, on Newlyn Downs, with only a slight frost on the road. The hind wheels stopped turning and there was a terrific noise but, luckily, the set stayed straight and no other inconvenience was caused.

As might be expected, the set also slipped away on the notorious Rejerrah Hill on its way to Newquay. Only the machine was in tow at the time and the steersman managed to "catch the bank" and stop without damage. On another day — or rather a wet windy night — Bill Annear was bringing the set up this hill and passed the same tree three times, the engine having got its load up so far and then slipped back again and again!

By this time the motion was also well worn and it was frequently necessary to shim up the gudgeon in the crosshead. A Tate & Lyle treacle tin cut up in strips served the purpose admirably. The crosshead slides were also severely worn and the guide bars were never right. Each side of the firebox was cracked above the first row of stays and had been welded on site at Newlyn Downs. The paintwork was also bad and whilst in this ownership the whole engine was painted and lined out by Baden Powell in his spare time. This action considerably improved the engine's looks, and the firm's name was painted on a piece of tinplate nailed to the toolbox. The cast iron stack had burned so thin that it cracked about 3 inches above the bottom flange, and one day fell off the engine on the road to Cubert. It was replaced the same day by Argalls of Truro. Nevertheless, the old engine was still capable of a full season's work, and indeed worked a season of 212 days during one of these years. Whilst in this ownership it was never given a name.

One day while threshing near Treamble, the engine was ordered to help load an excavator for the army at Penhale. The men on the threshing job had to wait while the engine went to do the job. During this operation the mouth of the hitch on the perch got broken, the machine being subsequently loaded without the help of the engine!

Around 1945 the firm purchased a new David Brown tractor but did not use it for nearly a year. The "War—Ag" set was also purchased and towards the end of the 1945/6 season the David Brown started to be used in place of the engine, which was put back in the yard. It lay idle until purchased by a young enthusiast in the autumn of 1948. The last man to drive it in commercial threshing was Will Annear and the last farm threshed was Ventonteague, Journey's End, near Carland Cross.

So ended a continuous working life of thirty eight years and fortunately for us, Paul Clark who had worked with it for most of these years, so loved his engine that he refused to sell it for scrap, in spite of a higher offer by Messrs. Benny of Pool, near Redruth. After purchase by M. Roberts of Carland Cross, who executed some minor repairs, it was steamed out of the parish and parked beside an old burrow in sight of the A30 road near Carland Cross in St. Erme parish. Since then it has been preserved, the only engine based in Perranzabuloe so to be.

The Sandpits

"For more than seventy years, farmers within several miles of Perran Sands have been in the habit of resorting to these vast mounds of sand which they use as manure........." West Briton, 1st June 1832.

In the age of the traction engine, therefore, it is not surprising that thousands of tons of sand were "exported" from the parish with the aid of engines instead of horses. Most of the sand came from the sand pit situated just off the road, at the bottom of Budnick Hill, near the left hand corner into Perranporth. Another was situated at Gear Sands, about a quarter of a mile or more across the dunes from the corner at Tollgate, near the top of the same hill.

When the first loads of sand were pulled away by an engine is not known, but it is certain that sand was being pulled before the turn of the century. This traffic had become so much of a nuisance by 1902 that complaints about owners of engines loading on the Highway at Ponsmere, "occupying half the road for two and a half hours", were brought to the notice of Truro Rural District Council and the local Police. As a consequence, the dumping of sand by the side of the road was prohibited under County Council By-Laws. The prohibition does not seem to have been enforced for long as sand was still being dumped by the side of the road as recently as the nineteen thirties.

When the sand was to be brought from some distance inside the pit, it was transported to the road with horses and carts, carrying a ton at a time, on the evening before the engine was due. Twenty one loads would be brought out, (tipped in distinct piles by the side of the road), as good measure for the traction trucks which usually carried 10 tons each in trains of two or three. As recently as the mid thirties the price charged for bringing the sand from pit to road was 8 shillings (40p), for the twenty one cartfuls. This price included the labour of two men to load the carts as well as the hiring charge. All loading both in the pit and from the road was done by hand.

A road into the bottom pit was made from spar stone brought down from Newlyn parish when Newlyn Downs were being broken in during the early part of the century. This allowed the engines to drive into the pit with their trucks, turn around and load direct. "When the sand was wet it was alright, but if dry very difficult".

The trucks were pushed back with a pole against the sandhill to facilitate loading. At another time, an anchorage was placed in the pit so that the engines could pull the trucks in with their wire rope around a snatch block and out again by a direct pull on the rope. This method allowed the engine to stand on the hard road at all times. At times there would be up to three engines there at one time, and on one occasion an engineman lost patience while waiting for the train in the pit to load and in an attempt to pass it, crushed the bank of the river, consequently ending up in the river with his engine! At other times there would be up to five engines a day pulling from the pit at Gear Sands where loading was direct to the trucks.

When pulling from the bottom pit, a stiff climb out of Perranporth is required on whatever road is taken. Frequently, in spite of this, the bigger engines of eight horse power or more hauled two or three trucks of sand up these hills at the same time. The sand was taken right across the county by some contractors (even as far afield as Feock by Messrs. Mitchell Bros.) Other contractors who are known to have pulled sand away were, The Tregony Co.; Powell of Summercourt; Simeon Rose, Mitchell; Obed Lanyon (and later his nephews Howard and John Henry); G. Carlyon, Mount Hawke; S. Carlyon, St. Day and W. Carlyon, Greenbottom; Simmonds Hodge and Messrs. Parsons, St. Columb; (Obed Lanyon and C.H. Carlyon were pulling sand well before the turn of the century). No doubt there were many more too.

The practice of pulling three trucks up the steep hills resulted in a fatal accident in 1902. Mr. J. Knight, a grocer in Perranporth, was walking with a horse up Liskey Hill, about thirty yards behind Messrs. Mitchell's engine and train loaded with sand, when, near the railway bridge, the coupling and chains of the last two trucks gave way. "The two trucks came down the hill at a terrific rate, the corner of the last one striking Knight between the eyes", causing instant death. The horse escaped injury, having become frightened by the approach of the trucks it turned and bolted down the hill. Similar practices also played havoc with the roads, which were deeply rutted and it was not uncommon for a representative of the County Surveyor to be seen hiding behind a hedge, checking for road damage as the engine passed. In latter years, by which time single wagons were usually hauled to the top before being made up in trains, the drivers hauled two trucks if they could get away with it.

During the thirties, much of the pulling from the bottom pit was stopped for a few years as so much sand had been removed. Sand is still pulled today but not with traction engines.

Messrs Parsons (St. Columb). Burrell 3816 railing narrow gauge locomotive "Dolcoath" after bringing it to Treamble Quarry from Dolcoath mine, Camborne.
Photo courtesy Mrs. M. Kitto

Other Engines

A Fowler road loco often seen on the roads within the parish was number 11441, called "Millenium King", and owned by G.A. Carlyon of Mount Hawke. This was a seven horsepower engine used mainly for haulage but quite often employed crushing stone. The

Fowler road locomotive 11441 "Millenium King", owned by G. Carlyon, contractor, Mount Hawke. This engine was frequently used in Perranzabuloe especially for crushing stone, near Perran Round. "Neddy" Mewton with cap.

Photo from the collection of D.E. Hattam

Council depot was situated near Perran Round and this engine was employed there on this type of work, especially in the 1920's. The stone was brought from Budnick Quarry, now filled in.

Another well-remembered engine, "King of the Road" driven by an equally remembered Jack King, often came to the area on hire from its owners Simmons Hodge & Co. of Truro. This was Burrell 2897, an eight horsepower traction. On one occasion this engine was used to do some rolling in the repair of a private road. The foreman expressed his satisfaction to King but the driver was not satisfied with his job. There was a deep ditch beside the road and a small width at the edge of the road that had not been rolled. It was not until after he had rolled the edge with the inner side of the wheel, the outside being in the air as it were, that the job was pronounced complete.

A Foden steamwagon is reputed to have been employed at Treamble Quarry. The quarry and wagon were owned by Steve Clark and the wagon was known as "Black Bess" on account of the smoke and fire that belched from it when pulling hard. It is reputed to have been, by this time, half cement, it having been used to stop up so many holes.

Several portable engines worked in scattered parts at various times. Those employed on bigger jobs are remembered best. A large one was at work around the time of the first War at Wheal Anna, sawing wood. Another engine also working on wood worked at the Treelight end of Mill Downs. Schoolboys would often spend time picking up small pieces of wood and feed it to the engine. During the first War another portable was employed at Chiverton Mine, assisting in work on the old mine burrows there.

Several people know of steam ploughing within the parish, some saw it taking place. One informant remembered as a boy, being told by old men, of steam ploughing with two engines in the field and the plough pulled by a rope. (This set was probably that of Sydney Davey already referred to). During the first War a ploughing set was employed at Ventonvaise Farm. At Goonhavern Moors, a single ploughing engine was used to plough the moors for stone. A single furrow ripping plough was used and after being drawn through by the engine, it was pulled back to the bottom of the moor by a horse.

A large number of fields, consisting of several acres (from 5 to 15), in the vicinity of Callestick Veor and Callestick Vean Farms, appears to have probably been broken in by steam ploughing. All the hedges run parallel, roughly NW—SE, and a wide lane, running almost parallel to the bottom of the valley, intersects the fields about halfway between the A3075 road and the stream in the valley. It appears that this lane was used as a common headland to the land each side. (The lane is now bordered by earth banks). Assuming steam was used here, the set would almost certainly have been Sydney Davey's.

A Garrett roller owned by R. Dingle & Son, contractors, at work on road widening near the top of Mount Hill. *From collection of D.E. Hattam*

Various road rollers worked here at times. Most of these were hired by the Truro Rural District Council. This work was in later years taken over by the County Council. Contractors in the early years would probably have been the Eddison Steam Rolling Co. of Dorchester, but from the early 20's Dingle & Sons of Stoke Climsland took over this work. Their last roller to be kept in the parish was Burrell 4012, now preserved and called "Ventongimps".

James Penna & Sons

James Penna came from Perranzabuloe parish, was brought up at the family home, Hendravossan, and then ran nearby Meadow Farm at Rose. A butcher by trade, Penna, one of a large family of children, was considered to be a leader, to whom they often turned for advice.

Some time before 1898, he moved to Trelowarren Home Farm in Mawgan parish, which he farmed. From there he also operated as a butcher and in addition, set up as a threshing contractor at about this time. It is believed that his first engine was a portable but nothing is known of it. He had a new 4′ 6″ Marshall machine in 1898 and in 1904 purchased a new 6 H.P. Burrell single crank compound traction engine. This engine is now preserved and was in fact, the founder engine of the West of England Steam Engine Society. As therefore there is a link with Perranzabuloe and preservation, albeit a small one, some notes are appended to conclude this work.

The engine, Burrell 2662, was new in June 1904. James Penna was proud of his engine, which had been built to his specification. This had included 6′ diameter hind wheels with polished nickel hub caps

Burrell 2662, c.1910. James Penna (with basket) provides crib while the photographer takes this shot of the workers. The man next but one on his right is Jim Rowland who also came from Perranporth and at this time was firstman.

From the collection of D.E. Hattam

Copy

ALL COMMUNICATIONS TO BE ADDRESSED TO THE COMPANY.
TELEGRAMS: should reach the Works before 5PM and on Saturdays before NOON to ensure prompt attention.
A.B.C. CODE USED. TELEGRAPHIC ADDRESS: "BURRELL, THETFORD"

CHARLES BURRELL & SONS, LIMITED,
Engineers,
Manufacturers of
ROAD LOCOMOTIVES, ROAD ROLLERS,
TRACTION & PORTABLE ENGINES,
STEAM PLOUGHING & CULTIVATING MACHINERY,
THRASHING MACHINES, &C.
ESTABLISHED 1770

Saint Nicholas Works,
Thetford, Norfolk.

Name Plate J. Penna & Sons
Trelowarren. Mawgan

Delivered by From Cornwall Show 1904 or as soon after as possible.
Station Falmouth
Railway G.W.R.
Carriage to be paid by Senders.

Terms Net cash on delivery

Bought of Charles Burrell & Sons, Limited.

By Mr Jas Penna: Trelowarren Farm
Mawgan
Nr Helston

Description. One new Six Horse Power Traction Engine fitted with Patent Single Crank Compound cylinders, Hind Wheels 6'0" x 16" with 7/8 crossplates, Front wheels with 1" over tyre & polished nickeled caps, Sight feed lubricator fixed in view of driver, Pipe to lead water under engine, copper pipe to pump clack box, Copper pipes to injector, Governor pulley 2 flanges, seat for steersman hinge to fold back. Large size Penberthy Injector & Pump plunger joint at bottom. Frost pins with cotters, Flywheel brake as well as other. Copper top to Chimney, Guide pulleys with top & bottom rollers for Wire Rope, Brass bands on cylinder lagging, Polished safety valve lever, starting & live steam rods, Iron Tool box on side & wood box at back, Oil cups on countershaft & axle brackets, Oil door on R.H. splasher below the door. Also outfit as per catalogue. Price £461 nett (Say four hundred and sixty one pounds) Bore holes for fixing awning over driver

Messrs. C. B. & Sons, Ltd., to send man to unload and start Engine. No charge to be made for his time for the first week of 54 hours, custom[illegible] finding him board and lodgings, and paying rail fares and expenses out an[illegible] home; also his wages in addition after the first week.

Dated April 27th 1904

Witness:-

Signed James Penna

The original specification of Burrell 2662, new to James Penna in 1904.

Courtesy of C.P. Gregory

to both hind and front wheels. Copper pipes for water delivery and a large sized Penberthy injector were specified as was a copper top to the stack and brass bands for the cylinder lagging. An extra flange was fitted to the governor pulley and the small end of the pump connecting rod was jointed at the bottom of the hollow plunger, to reduce wear. There was a flywheel brake and top and bottom rollers for the rope guides. The safety valve lever, regulator and live steam rods were polished and the cylinder covers were also nickel plated. Several other small improvements were also incorporated and a wooden box mounted on the back of the tender supplemented the iron one on the side of the boiler. A brass plate, lettered James Penna & Sons, Trelowarren, Mawgan, was fastened to the latter box (he was not very popular with his landlord for giving this address as opposed to Trelowarren Home Farm and it was said the landlord was so annoyed that Penna was made to bring the engines and machines in via the back entrance of the estate as retribution — a very much longer way). The engine was supplied and exhibited at the Royal Cornwall Show held at Falmouth in that year and the total cost was £461 for cash. There can be no doubt that this was a very smart looking engine finished in Burrell's maroon livery. It was 17′ 6″ long, 7′ 3″ wide and the stack top stood over 11′ above the ground.

At the show James Penna was offered a Devonshire type engine, also exhibited on Burrell's stand but he declined to purchase it as he could not afford to pay for another new machine to go with it. Apparently Mr. Burrell offered credit facilities as an added temptation but Penna still remained prudent.

In 1907, however, he had enough money to purchase a new Devonshire type engine and new machine. This engine was number 2950, being smaller than his other engine, and of only five horsepower. Like the former engine this had been also built to his specification. Business was good and about 1912 he purchased a nearly new engine and machine from the Lyonesse Threshing Co., Manaccan. Although this engine, number 3420, was also of 6 H.P. it was bigger than either of the others. The sale of this engine is reputed to have been forced by the Bishop of Truro, who was displeased to find that his incumbent at Manaccan, the Rev. Ormond, was the silent partner in the Company. For this reason, this engine became known as the "Parson's" engine and is still referrred to as such by older people of that area.

Burrell 2950 "Devonshire" type traction engine, new to James Penna in 1907 and now preserved in Cheshire. The "Devonshire" type new to Warne Grubb & Co. in the same year was very similar. *Photo courtesy A. V.K. Thomas*

Penna had three sons, two of which were actively engaged in the threshing business. One of them Richard John, had a flair for engineering and was entrusted with the general maintainance of the gear. He was much respected by the work's fitter from Burrell & Sons who came regularly to check over the engines. (A man from Marshalls also used to come to check the machines). It is believed that one of the other sons James Henry drove the first engine for a while when it was new.

With three sets at work several parishes could be covered, in fact, seven were worked; Mawgan, Sithney, St. Anthony, St. Martin, Manaccan, Breage and Wendron. Another engine, a Marshall, was added to the others at a later date but details are unknown.

In the first World War, under the Government Emergency Orders, farmers were expected to plough more acreage to produce food. Rather than plough up some of his best grazing land, James

Penna ploughed up the lawns of Trelowarren and had to face the landlord's fury by being given notice to quit! He then bought Godolphin Manor in Breage parish and spent three days and nights shifting all the farm gear, using his own engines and wagons. One of the drivers became so tired that he went to sleep on his engine, falling back into the coal bunker, on the road between Helston Hospital and Melrose!

Years ago, the writer had the good fortune to meet some of the men that worked for the firm and many interesting anecdotes were related. A few will suffice here.

Late one afternoon, one of the drivers was taking the machine from Tregedick to Teresa Farm, Porthleven. He had accidentally let the water level get too low with the result that the fusible plug started to weep. As he had no water in the tank he kept going in an effort to reach the horse trough at Garras. Eventually it was reached, by which time the fire was out and steam pressure almost gone. As soon as possible they dropped the ashpan and pulled out the firebars. The driver went up into the hot firebox "boiling hot and sweating", removed the plug and fitted another. They were then able to fill the boiler from the trough, put back the firebars and ashpan and relight the fire. It was the small hours of the morning when they arrived at Teresa Farm, to find everyone had gone to bed and no-one would answer the door.

Another time, during a very hard part of winter, he was proceeding to Lezra Farm and whilst going down over Lezra Hill the engine began to slip away on ice. He had no frost pins in at the time, much to the subsequent annoyance of Mr. Penna, and the engine eventually ended up in the hedge, causing damage to the smoke box and fracturing a corner of the perch bracket.

On one occasion, after threshing had been completed at Tranno Farm, Breage, someone bet the driver a pound note "that he couldn't go up there with the engine and machine and out through the gate at the top". To get to this top entrance it was necessary to ascend the steep side of the field in which they had been threshing. The wager was accepted and all the spuds put on, ten on each wheel. The greater part of the hill had been climbed, when a couple of spuds came off at the steepest part quite near the top. "I could feel the hill coming towards me and she digged in, threw her head up in the air until she was resting on the tender and I holding her on the reversing handle!" All hands now came to assist and trigged the

Burrell 2662, new to James Penna in 1904. This engine became the first engine bought for preservation in Cornwall. Seen here, c.1952, in the ownership of Charles Gregory, President of the West of England Steam Engine Society.

Photo from the owner's collection

engine up, put the locking chains on the machine and after all the blocking was secure, the front of the engine was lowered on the reversing lever! The engine then roped itself and the machine to the top of the field.

At a more leisurely period, this driver picked out his name, with the point of a penknife, on one of the brass boiler bands and it can still be seen to this day.

For some reason James Penna did not insure his boilers (this was not compulsory until 1936) but he did insist on keeping the engines extremely well maintained. A fitter employed by Burrells visited the concern each year and overhauled the engines as necessary.

He died in 1926 and the business was run by his two sons until 1936. J.H. Penna died in 1935. Shortly after this time all the sets were sold at Helston. Burrell 2662 was purchased by a Mr. Berryman for his son-in-law, Mr. J. Abell of Ludgvan; Burrell 2950 was purchased by a grandson of James Penna, Mr. W.S. Richards, St. Martin and Burrell 3240 went out of the County to a Mr. Ellacott of Yealmbridge.

The two former engines are now preserved and it is no doubt due in no small part, to the way in which these engines were so well maintained in their hardworking days — a credit to their owner.

APPENDIX

Summary of Engines Based or Regularly Worked in Perranzabuloe

Builder	Works No.	Year Built	N.H.P.	Cylinder Arrangement	Type
AVELING	3706	1896	8	Compound	Rd. Loco
& PORTER	4924	1901	8	Compound	Rd. Loco
	5052	1902	8	Compound	Rd. Loco
BURRELL	2464	1902	8	S.C. Comp	Rd. Loco
	2944	1907	5	S.C. Comp	Traction
	2897	1907	8	S.C. Comp	Traction
	3034	1908	6	S.C. Comp	Traction
	3500	1913	6	S.C. Comp	Traction
	4012	1925	—	Single	10 Ton Road Roller
FOWLER	1187	1869	12	Single	Ploughing
	1188	1869	12	Single	Ploughing
	5483	1887	8	Compound	Rd. Loco
	11249	1907	7	Single	Traction
	11441	1909	7	Compound	Rd. Loco
GARRETT	?	?	—	Single	10 Ton Road Roller
MARSHALL	5981	1878	5	Single	Portable

Engines Owned by J. Penna, Mawgan

BURRELL	2662	1904	6	S.C. Comp	Traction
	2950	1907	5	S.C. Comp	Traction
	3240	1910	6	S.C. Comp	Traction

Name (if any)	Reg. No.	Owner	Notes
None	None	Nobel's Explosives	Scrapped
None	None	Nobel's Explosives	Scrapped
None	None	Nobel's Explosives	Scrapped
None	None	Munitions Factory	Scrapped
"Little Engine"	AF 3579	Warne Grubb & Co.	Scrapped
"King of the Road"	AF 3332	Simmonds Hodge & Co.	Scrapped
"Big Engine"		J.P. Watts	Preserved
	AF 3580	Warne Grubb & Co.	
		Clarke & Nicholls	
None	AF 3341	A. Sobey	Scrapped C.1950
None	RL 1254	R. Dingle & Son	Preserved
None	None	J.S. Davey	Scrapped C.1950
None	None	J.S. Davey	Scrapped
"Gamma"	None	F. Barrett	Scrapped C.1932
None	AF 3581	Warne Grubb & Co.	Scrapped C.1947
"Millenium King"		G.A. Carlyon	Scrapped
?	?	R. Dingle & Son	Scrapped
—	—	R.J. Grubb	Scrapped
"Maudie"	AF 3531	J. Penna (Mawgan)	Preserved
None	AF 3532	J. Penna (Mawgan)	Preserved
"Parson's Engine"	AF 3533	J. Penna (Mawgan)	Scrapped

Acknowledgements

J. Sydney Davey:

West Briton; Royal Cornwall Gazette; Western Morning News; County Record Office; H.L. Douch, R.I.C; A. Duke, R.L.S; T. Knight; Redruth Ref. Library; Bibliotheca Cornubiensis.
A visit to Bochym — Thomas.
Annalls of a county town — Mitchell.
Road Loco Society Portfolios — Steaming Vol.
Steam Locomotion on Common Roads by J. Head, 1872.
Illustration — based on a line drawing after J. Head.

Agra Bank Mining Co:

R.G. Pratt, R.L.S; West Briton; Royal Cornwall Gazette; Mining Journal 1872; Mining Magazine Vol. 1 1872; Engineering; The Engineer.

Richard Joe Grubb:

W. Annear; M. Gifford; W. Green; E. Hosking; R. Mitchell; the late E. Pascoe; J. Payne; B. Powell; R. Salmon; P. Thomas; H. Tremewan and Mrs. Solway, all of Perranzabuloe parish.
K. Eley and F. Ennor, St. Agnes parish.
Photograph — courtesy Benney, Camborne.

Nobel's Explosives Factory:

W. Barrett; B. Green; W. Lavin; B. Mitchell; R. Salmon and J.P. Tredinnick, all of Perranzabuloe. K. Eley and F. Ennor, St. Agnes. Minutes C.C.C. Highways, bridges & main roads committee. Royal Cornwall Gazette 1895—1905 inclusive. Road Loco Society Portfolios.

Simeon Penna:

W. Annear; J. Davis; E. Hosking; S. Glanville; W. Green; M. Penna; S. Penna; W.J. Penna; B. Powell and Mrs. Solway, all of Perranzabuloe. Mrs. M. Kitto, Praa Sands. W.S. Richards, Mawgan.

Warne Grubb & Co:

Photographs — courtesy Mrs. P. Clark, Goonhavern and Mrs. O. Stephens, St. Agnes.
W. Annear; M. Gifford; R. Grubb; E. Hosking; E. Jewel; D. Juleff; R. Mitchell; J. Nicholls; J. Payne; B. Powell; R. Salmon; P. Thomas and Mrs. Solway, all of Perranzabuloe parish. K. Eley; F. Ennor; E. Jennings; H. Penrose and Mrs. M. Tremain, all of St. Agnes. M. Watts, St. Minver.
A. Duke, R.L.S; Burrell & Fowler catalogues.

J.P. Watts:

J. Davis; E. Hosking; M. Penna; S. Penna; W.J. Penna; P. Thomas and Mrs. M. Morgan, all of Perranzabuloe parish.
D. Watts; M. Watts; Mrs. D. Lander and Mrs. M. Veal, all of St. Minver. Mrs. M. Kitto, Praa Sands. B.D. Camp, Modbury.
A. Duke, R.L.S.

Arthur Sobey:

W. Annear; W. Barrett; W. Green; E. Hosking; J. Nicholls; J. Payne; M. Penna; S. Penna; B. Powell; R. Salmon; P. Thomas; G. Tonkin; J.P. Tredinnick; Mrs. E. Coward and Miss D. Westlake, all of Perranzabuloe. K. Eley and F. Ennor, St. Agnes. B. Hosking, St. Allen. G. Hawkey, Crantock.
A. Duke, R.L.S.
Photograph — courtesy Miss D. Westlake.

The Munitions Factory

W. Barrett; Miss E. Batten; B. Green; W. Lavin; B. Mitchell; J. Payne; B. Powell; W.J. Penna; H. Tremewan; W. Tregonning and A. Westcott, Perranzabuloe. K. Eley, Mithian.

F. Barrett:

W. Barrett, Perranporth. J.H. Trounson; A. Duke, R.L.S.
Minutes Cornwall C.C; Road Loco Society Portfolios.
Photograph — courtesy J.H. Trounson.

Clark & Nicholls:

W. Annear; E. Hosking and J. Nicholls, Perranzabuloe.
M. Roberts, St. Erme.

The Sandpits:

Bob Mitchell; J. Nicholls; W.J. Penna; B. Powell; R. Salmon and W. Tregonning, Perranzabuloe. K. Eley, Mithian. J. Radliffe, Cubert. B. Hosking, St. Allen. Royal Cornwall Gazette; West Briton.

Other Enemies:

K. Eley, Mithian. J. King, St. Agnes. Mrs. M. Kitto, Praa Sands. T. Mitchell; Mrs. M. Morgan; J. Payne; M. Penna; W.J. Penna; Mrs. O. Piper; B. Powell; P. Thomas; W. Tregonning and J. Trevail, all of Perranzabuloe. D. Solomon, St. Erme.

James Penna & Sons:

M. Penna and W.J. Penna, Perranzabuloe. W.S. Richards, Mawgan. The late J. Ferris. The late J. Williams, Nancegollan.

Index

Other titles available from

DYLLANSOW TRURAN

CORNISH BEDSIDE BOOK NO.1.

John Keast

Place-names, customs, dialects, remedies and recipes are nicely balanced by short essays and stories by early travellers, historical events and extracts from diaries. A throughly entertaining book.

Card Covers 9506431 8 1

VISITORS TO CORNWALL

Ida Procter

As J.C. Trewin says in his Introduction- 'Ida Procter can recreate her guests most surely... Everyone is here; Francis Kilvert for example, Ceila Fiennes is here - at St. Austell. Nearly two centuries later the young Beatrix Potter was arriving by train at Falmouth.

It is an enchanting book.'

Paperback 0 907566 27 8
Hardback 0 907566 26 X

HOW LONG IS FOREVER *Memories of a Cornish Maid*

Ethelwyn Watts

Born in the changing years of the decade that followed the First World War, the author describes her home parish as 'perhaps the most Cornish part of Cornwall' and she paints a word picture of a magical land.

Card Covers 0 907566 30 8
Hardback 0 907566 89 8

THE BOUNCING HILLS *Dialect Tales and Light Verse*

Jack Clemo

Jack Clemo says 'I contributed many dialect tales to Cornish Almanacks before the war capturing the lighter side of Clay country village life as it was 50 years ago. I have chosen eight of these stories, and have added a selection of my comic verse (not in dialect, but with a Cornish flavour). There are about 20 short poems, mostly written for or about children'

This is a book from a Cornish literary giant shedding a new light on Cornwall's own blind poet.

Card Covers 0 907566 38 3
Hard Covers 0 907566 39 1

THE HISTORY OF FALMOUTH

Dr. James Whetter

The author's considerable knowledge of life in the 17th Century Cornwall provides much new information about the origins of the old town - its growth over three and a half centuries and its social, cultural and religious history.

Hard Covers 0 907566 01 4
Card covers 0 907566 02 2

CARN BREA

Michael Tangye

A book on Redruth, Illogan and Camborne's dominating feature.

Michael Tangye's expert account will meet the needs of students of all ages and all of those who are curious about 'the Carn'. Illustrated by the author.

Card Covers 0 907566 11 1

CORNISH EPITAPHS Vol.1.

John Keast

Records, humour, pathos AND sermons in stone! Beautifully illustrated with photographs, line-drawings, etc. This is volume one of a two-volume work.

Card Covers 0 907566 15 4

THE LIFE OF PASSMORE EDWARDS

R.S. Best

(With an essay on the life of Sylvanus Trevail
by Peter Laws)

Rising from rags to riches, he endowed nearly 70 institutions - libraries, art galleries, hospitals, museums and an orphanage etc. throughout Britain between the years of 1890 and 1904. The first substantial biography of the great Cornish Benefactor.

Hard Covers 0 907566 23 5
Card Covers 0 907566 18 9

THE SAINTS OF CORNWALL

Catherine Rachel John

A comprehensive account of the many saints of Cornwall - dedications ranging from early Celtic to modern times. Miss John has built on the researches of her father, (Donald Attwater) Canon Doble, and others. Illustrated with 40 photos aod line-drawings.

Linen Covers 0 907566 14 6
Hard Covers 0 907566 13 8

MEMORIES OF OLD PENRYN

Florence Rapson

(Edited by Rita Tregellas Pope)

A picture of a way of life of which only nostalgic memories remain.

0 907566 42 1

KERNEWEK MAR PLEK!

The newest course in Cornish that has already revolutionised Cornish Language teaching

KERNEWEK MAR PLEK - First Course

Crysten Fudge and Graham Sandercock

9506431 2 2

KERNEWEK MAR PLEK - Second Course

Crysten Fudge

9506431 5 7

Cassettes are available to convert this series into a home tuition course for students of all ages.

CORNISH SIMPLIFIED

Caradar (A.T.D. Smith)

Edited by Talek (E.G.R. Hooper)

Still the most complete reference book of Cornish grammar.

0 907566 09

THE LIFE OF CORNISH

Crysten Fudge

The story of our language to the 17th Century. An immunsely scholarly and readable work. Attractively illustrated. A fascinating story told with the insight and understanding of a foremost Celtic scholar.

Card Cover 0 907566 20 0
Hard Cover 0 907566 24 3

CORNISH NAMES FOR CORNISH HOMES

Crysten Fudge, M.A.

A name-it-in-Cornish handbook, delightfully illustrated by Laura Rowe.

9506431 0 6

NAMES FOR THE CORNISH

300 Christian names for your children - all of them Cornish. With derivations and translations.

0 907566 03 0

A HANDBOOK OF CORNISH SURNAMES

G. Pawley White

Dr. A.L.Rowse points out in his foreword that the study of Cornish surnames is an essential work in the elucidation of the distant Cornish identity.

9506431 9 X

A HISTORY OF SOUTH CROFTY MINE

J. Allen Buckley

The history of this Queen of Cornish Mines, told by a working miner extraordinary. This book with photographs, illustrations and maps gives a detailed history of 300 years of working the property, lode by lode and level by level.

Hard Covers 0 907566 17 0

ST. IVES MINING DISTRICT - Vol.1

Cyril Noall

In this book the whole fascinating story of mining around St. Ives is related with a wealth of detail and illustrations with numerous photographs, sections and plans. The Author, historian and Bard of the Cornish Gorsedd has already written many learned histories of Cornish mines, customs and events.

Paper Back 0 907566 33 2
Hard Back 0 907566 32 4

THE HISTORY OF THE CORNISH COPPER COMPANY

W.H. Pascoe

This book describes the History of an important undertaking in West Cornwall throughout the years of Britain's Industrial Revolution. It is the story of the only successful copper smelting firm in the whole of Cornwall, a firm also in the forefront of the development of the internationally famous Cornish Steam Engine.

Paper Back 0 907566 35 9
Hard Back 0 907566 34 0

THE KING OF MID-CORNWALL

The Life of Joseph Thomas Treffry

John Keast

Joseph Thomas Treffry (1792-1890) has been aptly described as 'A captain of industry, master of his own economic fate, an assertive individualist soul who pioneered in the development of transportation to facilitate the exploitation of his mines and the marketing of the ores they produced!

Paper Back 0 907566 29 4
Hard Back 0 907566 19 7

STATELY HOMES IN AND AROUND TRURO

Rex Barratt

His book misses nothing of the gems of a favoured area - from ancient Polwhele and Killagordon (to which film star Kathleen Hepburn was a frequent visitor) to Tullimar - where Kilvert, of diary fame, spent a holiday in 1870.

Card Covers 9506431 6 5

ST. IVES HERITAGE

Lena & Donald Bray

Recollections and records of St. Ives, Carbis Bay and Lelant.

'This patient, well researched and loving look at the history, customs and general lifestyle of this jewel of the North Cornish Coast makes this an unusually endearing narrative' WESTERN MORNING NEWS.

Hard Covers 0 907566 07 3
Card Covers 0 907566 08 1

JOHN KNILL - HIS LIFE AND TIMES

Beryl James

'John Knill, Mayor of St. Ives, customs official and adventurer extraordinary, was a remarkable son of the rumbustious 18th century... The life-style of this lawyer-cum-pursuer of smugglers makes a scenario fit for any Holywood epic.

Card Covers 0 907566 06 5

CORNISH SAYINGS, SUPERSTITIONS AND REMEDIES

Kathleen Hawke

'What makes all Mrs. Hawke's work on dialect particularly interesting is that she has lived for many years in both mid-Cornwall and west-Cornwall' K.C. PHILLIPS.

0 907566 04 9

CORNISH QUIZ

R.S. Best

Over 500 questions and answers on Cornwall and things Cornish. Illustrated by C.M. Pellowe.

9506431 4 9

MEMORIES OF A TRURONIAN IN WAR AND PEACE

Rex Barratt

The picture of the city that emerges, how it adapted its ways of life to war, the excitement of Royal Visits, life in church and school, politics, music and drama, shopkeepers and local characters, are all faithfully recorded and illustrated.

Card Covers 0 907566 00 6